THE CREATIVE POWER OF CHANCE

THE CREATIVE POWER OF CHANCE

RÉMY LESTIENNE

Translated from the French by
E. C. Neher

University of Illinois Press
Urbana and Chicago

Published by the University of Illinois Press
by agreement with Éditions La Découverte
Manufactured in the United States of America
1 2 3 4 5 C P 5 4 3 2 1
This book is printed on acid-free paper.

Library of Congress Cataloging-in-Publication Data
Lestienne, Rémy.
[Hasard createur. English]
The creative power of chance / Rémy Lestienne ;
translated from the French by E. C. Neher.
p. cm.
Includes bibliographical references and index.
ISBN 0-252-02386-2 (cloth). —
ISBN 0-252-06686-3 (pbk.)
1. Chance. I. Title.
BC141.L4713 1998
123'.3—dc21 97-45250
CIP

To she who was met by chance
and who became my wife

■

I dedicate this book also to Thomas Jefferson, the third president of United States, and to his wife, Martha. In a handwritten note that has been preserved, Martha copied this sentence, quoted from Laurence Sterne's *Tristram Shandy:* "Time wastes too fast: every letter I trace tells me with what rapidity life follows my pen. The days and hours of it are flying over our heads like clouds of a windy day never to return—more everything presses on . . ."

On the same sheet of paper, Thomas Jefferson completed the quotation: "And every time I kiss thy hand to bid adieu, every absence which follows it are preludes to that eternal separation which we are shortly to make!"

CONTENTS

THE CREATIVE POWER OF CHANCE

INTRODUCTION

When he received the Nobel Prize in 1984, the physicist Carlo Rubbia declared that in his opinion, future generations will respect ours not for our industrial power or technical achievements but for the significance of the questions we asked about the workings of the world and for the bold originality of the experiments we devised in attempting to answer those questions. The ancient philosophers' deliberations were certainly of great relevance, but never before has science joined with philosophy in addressing with such enthusiasm questions about the ultimate nature of matter, its origins, its destiny, and our place in the universe.

Chance has made us members of this generation. Despite their brand-new and still incomplete understanding, researchers of our generation have obstinately sought answers to questions that do not differ greatly from those asked by their predecessors. How can we be sure that our senses are not deceiving us about the reality of the world? What is the true nature of the perceptions that most intimately affect our lives—the passage of time, for example? When I wrote my first book, *The Children of Time,*[1] I noted that chance continually intervenes in questions relating to temporal irreversibility. This seems strange to a physicist trained in the tradition of Galileo, Newton, and Einstein. The notions of chance and time resemble each other in more ways than one. Both lie at the root of our perception of existence. Both are simultaneously essential and difficult to discern or to define operationally. Neither allows itself to be confined within the framework of the physical theories of nature. Finally, both play a role in the natural sciences—a discrete although fundamental role in physics and a much more striking role in the life sciences. A serious study of these two notions today demands that both the physical sciences and the biological sciences be rigorously interrogated in turn—rigorously, because the interrogation, whether of a philosophical, psychological, or religious nature, can no longer be based on approximations, premature enthusiasms, or chancy constructions that would distort the value of the scientific method's observations.

Since I first approached the study of cerebral mechanisms, I have begun to understand the extent to which our brains' imperious demands for a coherent vision of the world can mislead us and explain away the diversity of views presented as factual reality, when they are actually no more than proof of our minds' ability to fabricate personal visions of the material universe. To satisfy this need for coherence, the brain often suppresses any of the elementary perceptions it receives that contradict the character of its system. Moderation should therefore rule our approach, but that moderation should not be mistaken for indifference. To speak yesterday of time, to speak today of chance, is to speak in both cases of ourselves. Marcel Proust would not contradict me on this point, although the approach I have adopted certainly differs from his. Doubting the merits of introspection, I have come to believe that it is better to rely on the scientific method, which is external and intersubjective. Its contribution to understanding the human spirit is of course limited, but it is still the surest guide at our disposal.

HEAT: A WORLD BATHED IN A SEA OF CHANCE

Heat is the first thing we must examine to understand the nature of chance. In the universe we inhabit, every substance, material or not, is associated with a temperature. What is heat? Since the end of the last century, physicists have responded that it is the manifestation at the macroscopic scale of the disordered motion of molecules. Disorder is therefore everywhere. It is around us and in us, in the form of that incessant vibration of matter's ultimate components. Since the end of the last century, it has also been known that this disorder is indispensable to the manifestation of temporal irreversibility, the "arrow of time." If the molecular vibration is channeled—for example, by giving all molecules the same orientation—you have not heat but motion, kinetic energy, which is reversible. A ball can be bounced so that it returns to its point of origin. Heat flow, however, cannot be reversed. There is therefore a difference in nature between order and disorder at the molecular level. The disorder of motion manifested in heat contains an intrinsic irreversibility. To attempt to generalize this observation, I propose calling each of the manifestations of chance marked by a congenital link between disorder and irreversibility "natural chance." This chance is distinct from mathematical chance, as in the chance of card games, which can produce the most astonishing hands, only to have them disappear and return anew if the cards are shuffled long enough.

This preliminary distinction may seem pointlessly to evoke a magic of chance. The recent development of deterministic chaos theories gives the impression that the mysteries of chance have been definitively unveiled, that chance is nothing but an illusion linked to the extraordinary variety of destinies common to physical systems with similar original conditions (what physicists call "instability in relation to initial conditions"). It is true that a physical system strictly obeying a deterministic law, bringing about changes thought to be reversible, can give rise to phenomena that appear to be random, such as tropical storms that may move from possible to actual at a single flap of a butterfly's wing. But here we must be precise. Deterministic chaos is in no way comparable to chance. In fact, they may even be opposites, since the former lacks the essential ingredient, the unpredictability *in principle,* that the true chance of statistical thermodynamics shares only with quantum mechanics. The fact of irreversibility is present in almost all situations envisaged by experimental science and the observation of nature, not only in the phenomena usually studied by thermodynamics or quantum mechanics. Is this not an indication that chance should be explicitly introduced in all the other branches of knowledge? The contemporary physicist Roger Penrose holds this view. He thinks that the future "theory of everything," if there is to be such a thing, will have to be a quantum theory of gravitation endowed with the two essential ingredients of unpredictability and irreversibility.

From another perspective, the goal of science is to unveil the hidden order of nature, and its progress to date testifies that nature does not consist just of nonsense or of an absence of causes. In the last century, physicists were already asking themselves how chance can engender order, and many biologists since Darwin have questioned how the impressive evolution of living beings, with the emergence of complexity and consciousness that seems to be associated with it, can be explained by the blind interplay of mutations.

We know the answer given by the partisans of the synthetic theory of evolution. It is natural selection that drives living beings toward ever greater complexity. But how then can one speak of progress? Is the progress of evolution continuous and ineluctable, or is it nothing but an illusion?

Darwin himself was not sure. "The naturalists," he wrote, "have not yet defined, in a way satisfying to all, that which can be understood as the 'progress of organization.'"[2] This question has been raised again in modern times by Motoo Kimura, among others. He demonstrated that in many instances, mutations and restructurings of genetic material are neutral with respect to evolutionary progress and therefore cannot explain it.[3] For their part, Daniel Brooks and E. O. Wiley propose interpreting the apparent progress of evolution as a simple effect of perspective.[4] According to them, the diversity of living systems increases in conformity with the law of increasing entropy, but because of the constraints of heredity, it does so much less than the number of theoretically imaginable variations would allow. In other words, the number of living organisms increases much more slowly than does the number of ecological niches made available by evolution, while the organisms incorporate and organize more and more matter. As Stephen J. Gould astutely noted, it would be wrong to represent the evolution of living beings as a tree thick with living branches. Most of those branches are dead limbs testifying to life forms that have disappeared forever.[5]

Darwin's disciples, disregarding his scruples, have militantly brandished the standard of Darwinism to support a lay explanation of the visible progress of evolution. The opposing camp, partisan to a theological explanation, was not fooled. It was precisely the fact that Darwin's evolutionary theory could not be reconciled with natural theology that made it unacceptable to many of his contemporaries. It was out of the question for them to admit that chance and the struggle for life might shape the destiny of the living world. And Darwin himself affirmed variation to be totally independent of selection. He took that independence to be the very definition of what he called natural variation. He thus instituted between the two orders of phenomena an almost total rupture that allowed no possibility for pairing the two aspects and subjecting them to some preestablished divine plan.

Darwin did not himself seek to offend believers, however, and he professed a prudent agnosticism in regard to religion. His proposal was at once more modest and more ambitious than a simple confrontation with theology. What he was in fact trying to show on the scientific level was that chance has an *objective* reality, or at least (and more precisely) that the chance in his theory merits consideration as having the same degree of reality as any other theoretical entity introduced in physical theory.

It is this proposition that I would like to reexamine today, armed with and enriched by all the findings from one and a half centuries of scientific research. This research, whether investigating physics and the immense distances opened up by the initial explosion of the universe, the physics of elementary particles and the chemistry of dissipative systems, or the blossoming of the tree of life inventing new species, has discovered that chance plays a preeminent role in all these elemental stages. Like a designer of fireworks who makes glittering enchantments erupt in the stillness of the night, chance creates—before our eyes and, as it were, through successive explosions—most of the marvelous constructions of nature. To crown all the extravagance of its displays, chance seems to have purposefully invented sexuality and its entangled lotteries to make of us, the latest arrivals in this amazing fresco, truly the Children of Chance.

PART 1

CLASSICAL CHANCE

A BRIEF HISTORY OF THE CONCEPT OF CHANCE

1

From antiquity until the nineteenth century, chance was associated mostly with gambling. Over the centuries, however, the notion of chance has come to be applied to other situations—in particular, to conflicting eyewitness testimony (especially in the courts) and to the tallying of the number of occurrences of a particular event among many possible events (the elaboration of statistical data). Each of these uses led to an expansion of the definition of chance, or rather to the development of concurrent interpretations of probability. All these premodern interpretations shared a common foundation, however. Their chance was only a word invented for convenience in summarizing a confused situation, not a real thing that kept the world from being purely deterministic. Chance was the veil behind which we hid our ignorance of an event's causes without challenging the existence of those causes.

FORTUNE AND CHANCE

The standard models of the chance of games are either shaking dice in a cup or shuffling a deck of cards. The motion of the dice in the leather cup and their erratic trajectory when thrown on the table are too complicated to be followed step by step. Similarly, the agile play of the shuffler's fingers through the cards is too rapid for all the changes in a given card's position to be followed. These actions result in situations that give all the appearance of being products of chance. But the players' actions do not really imply a discontinuity or break in causality.

Antiquity, the Middle Ages, the Renaissance, and the early modern period generally had a great repugnance for the most basic implication of chance, that of a complete absence of cause. These eras may be described as basically ultrarational since every event had to have an explanation, a reason, to be acceptable to the thinking person. Each event had to have a cause, although that cause could be mythological: the frightening sound of thunder is a battle between the gods, and so on. The triumph of the Christian religion in the West did not lessen this need for rationality—quite the contrary, in fact. Christianity stated that any event to which a reason could not be ascribed could occur only through divine intervention. Having to imitate the divine wisdom, humans could not allow their destinies to be determined by the vagaries of those factors that determine the outcomes of games. By tangling the clear thread of an orderly destiny, chance was suspected of being the cloak of the Prince of Darkness.

Antiquity did allow for the notion of contingency, of events that might or might not arise in the future, but with the notable exception of Lucretius, it did not get much involved in the objective absence of causes in contingent situations. Remember that for Plato, the primal substance of the world was chaos, but the demiurge intervened specifically to give it form. A first reservation on the necessity of the world was expressed by Aristotle. In his treatise *On Interpretation,* he discussed at length the notion of contingency, to which he attached great importance in terrestrial and sublunar events (as opposed to the necessary events of the celestial spheres beyond). In that text Aristotle was already saying that all possible outcomes of a contingent event do not necessarily have the same chance of being realized. Here as elsewhere, Aristotle showed himself to be astonishingly prescient (Popperian propensities bear a strange resemblance to Aristotelian potentialities, and for that matter to Heisenberg's potentia).[1]

Chance having no antecedent, it should not be surprising to see that the ancients varied in their efforts to assign it one. Thus, for the Greeks and Romans, the left and events arising from the left were interpreted first as happy omens and then later as indications of disasters to come. The truth is that the Romans had shrewdly observed that the left hand was the one that thieves used to lift jewels from passersby. The Latin word meaning left, *sinistra,* gives us the word *sinister,* applied to omens and calamitous events. But the Romans also used the word *fortuna,* which is the basis for our expression "good fortune," with a much happier connotation. Fortune is the chance that smiles on the well-born child, but anyone can occasionally claim its favors and seize its largess. It can also be called forth. Good fortune alone gave the peasant or the poorly born a chance to reverse the established order through, for example, an act of bravery in war and to acquire the titles that birth had denied. Although novels such as *Ben Hur* caricature the periods they describe, many commoners and slaves were able in this way to find redress from an unfavorable fate.

A FIRST STEP TOWARD THE TAMING OF CHANCE

The French word for chance, *hasard,* is of Arabic origin. We think that the crusaders became acquainted with it during their expeditions by playing dice, then in widespread use in the Middle East. The dice were called *az zahr,* which quickly became converted into the single word *hasard,* especially in relation to gambling.

A first mathematical approach to games of chance was attempted in the seventeenth century with an exchange of letters between Blaise Pascal and the chevalier de Méré, who was addicted to playing dice. On July 29, 1654, Pascal, ill and bedridden, wrote feverishly to the knight to tell him that he had solved the problem the latter had posed. The central notion he used in this first attempt to uncover a mathematical solution to the problem of chance was that of a *fair game;*[2] in other words, the player's decisions must be based on what is today called mathematical expectation or hope of gain. Thus it is not true that the first notion discerned in the theory of chance was that of probability. In Pascal's first discussion with the chevalier de Méré, probability was not specifically mentioned, and Pascal even remarked in the vaguest of terms that "chance is equal," meaning perhaps that he recognized only cases of equiprobabilities.

A new step was taken in game theory with the appearance in 1713 of Jakob Bernoulli's book *Ars conjectandi,* whose approach was already based much more on economics. Nikolaus Bernoulli, Jakob's nephew, was among the first to compose annuity tables. The two men also expanded the scope of questions linked to chance to an evaluation of the credibility of eyewitness testimony in the courts.

It was only at the end of this period, toward the middle of the eighteenth century, that the notion of probability would emerge as primary and be used as such in theories of chance.[3] When Laplace finally wrote his *Théorie analytique des probabilités* (Analytical theory of probabilities), he gave the place of honor to probability, in the classic form of a ratio between the number of favorable outcomes and the total number of possible outcomes, it being understood that the various outcomes considered were equally possible. Later this formulation was ridiculed by Poincaré, who said that it came down to defining the probable by the probable. Meanwhile, however, the field to which the calculation of probabilities could be applied was greatly expanded, and the suit that had been tailored for probability[4] was seen to have become much too tight. Research on the epistemological and axiomatic foundations of probability theory became imperative, independent even of the theory's economic applications, which had begun to appear.

STATISTICS AND INSURANCE

The origin of the practice of statistics is lost in the mists of time. Nonetheless, we can at least identify some clear traces of the notion's generalization over the course of the seventeenth century. At the outset the term was used in a sense quite different from the one used today (as its etymology suggests, *statistics* then referred to the "science of states"). It was not until about 1830 that the meaning given to this term evolved, first toward that of "the numerical science of society" and then, near the end of the century, toward it current meaning, the handling of large amounts of data and the art of prediction based on their study, whatever the object of that study.

In 1693 the British astronomer Edmund Halley (the same man who gave his name to the most celebrated comet in the solar system) published the first complete mortality tables making reference to the age of death.[5] At that time he introduced the concept of the survival curve and defined life expectancy. As Nikolaus Bernoulli had done based on other data, Halley calculated how much life insurance premiums should increase as a function of age for an equitable system of insurance. Both their works, however, were completely ignored by the professionals, for reasons that today seem fairly clear.

In the seventeenth century, insurance company practice was not much different from gambling. Insurance companies took care to give themselves comfortable margins of profit and had no interest in a better understanding of the risks they ran based on mathematical computations. In addition, in those uncertain times of piracy, epidemics, and brigandage, of great conflagrations and of war, they believed the variability of risks to be too great to be represented by the statisticians' well-balanced mathematical averages.

At first, as a function of their resemblance to gambling houses, insurance companies had a serious problem with respectability—all the more so because at that time, the practice of gambling was considered immoral or was even illegal. A papal decree in the thirteenth century (*Naviganti,* by Pope Gregory IX) formally interdicted the practice of maritime insurance in ordinary circumstances. In the middle of the sixteenth century, judges tried to develop the concept of an equitable contract based on a just ratio between the risks run and the premiums paid. But insurers, in a class with gamblers, paid dearly for the right to exercise their profession. This confusion between gambling and the gain associated with risk, on the one hand, and the prudent practice of insurance, on the other, continued for some time. Almost all the contracts used in the eighteenth century were more like the former than the latter. The enormous variety observed in natural risks at the time—for example, from fire or the uncertainties of maritime transportation—did not yet favor a rational approach to these problems. Finally, in England legal acts passed in 1764 and 1774 (the Gambling Act) based the distinction between insurance and gambling on legitimate interest. For example, the subscriber to life insurance had to prove a legitimate and proportionate interest in the life insured, without which the policy could be declared null and void. Life insurance nevertheless remained illegal in most European countries until the nineteenth century. Progressively, however, it was seen that, far from encouraging risk taking, insurance tended to diminish unbearable risks and to temper the injustices of destiny, and that it was perfectly compatible with the behavior appropriate to a responsible head of a household.

This was especially the case with life insurance. In 1762 the first insurance company founded on mortality tables, The Equitable Society for the Assurances of Lives, was created in London. For the first time, insurance premiums were modified as a function of the subscriber's age and their calculation was based on mortality tables. The company's initial prospectus insisted on its mutuality. Risk was diminished by being spread among all the insured. Benefits or eventual deficits would be shared. In fact, however, the founders, still fearing extravagant fluctuations, kept comfortable margins and ensured the rapid growth of their enterprise's profits. In addition, the practice of those first insurance professionals still exhibited the difficulty of truly understanding the relationship between probability and time. While the insurance promoters tended to identify time with an increase in uncertainty, whether because of the greater possibilities of an epidemic or because of unforeseen changes in the conditions of a voyage, probability theorists, to the contrary, knew that the regularities promised by the theory appeared only over time, from the patient accumulation of experiences.

The introduction of statistics into the practice of insurance marked an important turning point in the theoretical treatment of chance. Indeed, it originated the "frequentist" interpretation of probabilities, breaking with earlier approaches founded on games of chance (in which probability was estimated from the symmetry of the possible outcomes) or on subjective judgments, as with the reliability of eyewitness testimony.

The use of mortality tables in the evaluation of survival probabilities marked the first concrete example of a frequency approach to probability. Its origin can be seen in Jakob Bernoulli's distinction between a priori probabilities (in which symmetries bring about equiprobability) and a posteriori probabilities (when probabilities are calculated from statistics of results). Remarkably, Leibniz was already objecting that since statistics are necessarily limited, the choice of representing probabilities by one concrete series of results rather than by another is arbitrary. Bernoulli felt it necessary to respond that perhaps "nature followed the line of least resistance," which implies that all random situations probably have some underlying uniform distribution of probabilities. In the case of mortality tables, this claim proved difficult to translate into the theory, and the first treatises on the subject were grossly wrong, affirming against all evidence the equiprobability of the chances of death at any and every age!

MEASURING PROBABILITY

At the end of the eighteenth century, then, there were three ways to measure the probabilities of various possible outcomes in a situation of chance:

1. The examination of the physical symmetries present in a concrete problem often allowed equal probability to be attributed to each different possible outcome. Thus, the rotational symmetry of a cube led to attributing an equal probability of $1/6$ to the appearance of each of the numbers inscribed on the faces of a die.

2. Some, more sensitive to psychological matters, developed the subjective aspects of probability as intimate and eminently personal beliefs about the hopes and risks linked to different possible outcomes. This position, natural among men of law, is not the easiest to defend. How could one hope to construct laws from a theory founded on such a subjective definition of probabilities? How even to guarantee the intersubjectivity[6] of the debate on chance in this context, a minimal condition for it to be considered scientific?

3. Finally, a third approach appeared with the development of statistical methods. It based the evaluation of probability on patient observation of frequency. Trials of chance that could lead to several possible outcomes, such as a throw of the dice, were purposely repeated, or their natural repetition was awaited (e.g., for statistics on rainfall). Each observation enriched the statistics by one unit. For each event, the percentages of the different results observed determined, by definition, the probability of each outcome in the future (note that I am speaking of statistics for past events and of forecasting probability). This approach is derived naturally from the statistical practices concerning, for example, mortality tables and birthrates.

THE HIDDEN RULES OF CHANCE

The eighteenth-century men of science appear to have been fascinated by the idea that, beyond the fundamental disorder present at the occurrence of random events, those events seem to obey some higher order that forces them to manifest certain overall regularities. More precisely, the justification then invoked for the laws of chance insisted on the reality of the causes of the observed regularities. Probability was marked, therefore, by a Platonic vision of phenomena of chance. Why does the proportion of balls of a given color produced by repeated drawings of, for example, blue and red balls in a jar tend to reflect their proportion in the container? Because, it was then said, this proportion is the true cause of the observed regularity. Nonetheless, such a cause does not figure in the detailed actions that result in the drawing of one particular ball. There is no dynamic relationship between the contents of a jar and the sequence of mechanical steps of such a drawing. Another example, no less astonishing, is to be found in the regularity of death rates. That regularity, observed first in 1741 by the German pastor Johann Süssmilch and continuously by others ever since, allowed a sort of ideal curve of death risks to be defined. Thus it could be said, before the elaboration of a frequency interpretation of probabilities, that these empirical regularities encountered in the phenomena ruled by chance revealed the existence of a Platonic world of causes and reasons that in some way transcended the concrete events of daily life.

Today it could be said that these regularities are created by chance itself, by virtue of the law of large numbers. But that was an argument unacceptable to people in the eighteenth century, for whom the idea that order might arise from chance was anathema.

Strengthened by these observed regularities, theologians attempted to regain possession of chance and asserted that God holds it on a leash, so to speak, setting in advance a limit on the size of variations that the accumulated random variables can assume. The most admirable example of divine wisdom and the most convincing example of its intervention were thought to be found in the distribution of the sexes at birth. Did not the birthrate of boys surpass that of girls, in the proportion of twenty-one to twenty? Did not God thus foresee and compensate for the warring nature of adolescent boys, which is detrimental to their survival, in order to bring about a perfectly harmonious proportion between the sexes by marriageable age?

In this view of a controlled chance, the men of science found additional arguments to suggest that the introduction of probability was only an artifice or provisional expedient due to a temporary ignorance of the underlying causes of apparently random phenomena.

The nineteenth century reversed this view, tentatively at first, with Antoine Augustin Cournot's proposal of a categorical definition of events of chance. These events, he argued, were at the intersection of two chains of events, the chains linking events to each other through a series of causes and effects. Each chain, however, was totally independent of the other. The new point of view about chance was strengthened, in a manner as revolutionary as it was insidious, during a miraculous year: 1859. In a break with Laplacian epistemology, chance was evoked twice during that one year as an entirely separate element of a scientific explanation. First James Clerk Maxwell founded the mechanical statistics of gases by explaining their energetic properties as the disordered agitation of molecules, and a few months later Charles Darwin explained the evolution of living beings by invoking the chance of variations in an environment subject to natural selection. It was in that year, without any doubt, that the true scientific career of chance began.

1859: THE TWOFOLD INTELLECTUAL REVOLUTION OF DARWIN AND MAXWELL

2

The beginning of the second half of the nineteenth century was stamped with the imprints of iron and coal. Innovation and industrialization fed new faith in the progress of humankind while allowing Victorian England to assert its leadership over other nations. It was in an England at peace and sure of itself that a dual intellectual revolution took place in 1859. The explosive emergence that year of chance as an *explanatory element of phenomena,* in physics on the one hand and in biology on the other, constituted a true break with the ancient order, even if that break was not immediately recognized as such. The fact that Maxwell's hypothesis and the publication of Darwin's book dated from the same year is more than coincidence. It marked a coming of age of scientific thinking, henceforth confident enough in the explored field of causal determinations to launch an assault on other terrains. Before bringing on the actors, however, I should set the stage.

AN ERA ALIVE WITH INNOVATION

No other era—except perhaps our own, with its electronic and information revolution—has known such blossoming of technological breakthroughs and promise of social upheaval. While the railroad and the telegraph spun their webs, while the steam engine conquered the seas and invaded manufacturing, the incandescent lamp, the internal combustion engine, and the extraction of oil were invented in that single year of 1859.

It should not be surprising, then, that a wave of enthusiasm for progress swept through bourgeois society, which was becoming rich as never before and whose social customs were evolving rapidly. In France that year saw the completion of the first program of railroad construction, the principal engine of the economic expansion of the time. The wealthy British of Glasgow or Manchester began to discover the wonders of the Côte d'Azur. Railroads and telegraphs shrank the world. In 1858 the first transatlantic cable was brought on line, bringing the United States closer to Europe. The following year Ferdinand de Lesseps broke ground for the Suez Canal, bringing the East Indies closer to Europe. Already a telegraph network more than 10,000 kilometers long linked Paris with the provinces and allowed the government immediate and direct access to provincial prefects and notables, who until then had benefited from an insulating distance.

For the leisured class, daily life changed rapidly. Banks multiplied: the Crédit Industriel et Commercial opened in 1859, and the Crédit Lyonnais followed in 1863. The first department stores opened: Le Bon Marché and Le Louvre, followed by Le Printemps. Photographers' studios flourished in all the large cities. In Paris the Boulevard Sébastopol and the Gare de l'Est had just been opened, and the city was being transformed under the continuing direction of Prefect Haussmann. At the same time in London, excavation had begun for the first subway lines.

To some extent confidence in a better future inspired even the proletariat. Certainly the living conditions of the workers in the large cities remained miserable and in some ways even worsened over the years. The hourly wage rose, but more slowly than the cost of living. The meager incomes and trade wars in the business community led to a work force that included women and children. On the whole, however, the workers of 1859 lived better than did their counterparts of 1840, those indigent workers whose salaries had stagnated or dropped, who had only part-time work, and who suffered from endemic scarcities.

The industrial world was dominated by England, then at the height of its influence. At peace with the outside world since the end of the Crimean War, the aristocratic democracy of Queen Victoria allowed the English bourgeoisie to garner the benefits of its technological lead over the European continent.

In France 1859 was a rather tranquil year in the precarious calm characterizing the Second Empire. French troops were involved, it is true, in a short conflict in Italy.[1] Solferino's uncertain and dearly bought victory dictated a quick return to peace.

Prussia and the Austro-Hungarian Empire split their influence over Eastern Europe. In Prussia King Wilhelm I was still only the regent and had not yet called on the services of Bismarck, who was named prime minister in 1862. In Vienna in 1854, the marriage of Emperor Franz-Joseph to Elizabeth of Wittelsbach, called Sissi, dominated salon conversations, just as the marriage of the emperor to Eugénie de Montijo in France had done one year earlier. Habitués of those same salons, in whispered voices, spoke also of the latest scandalous books, such as *Madame Bovary* and *Les Fleurs du mal.*

America remained the new frontier. In 1859 James Buchanan presided over a union concerned first and foremost with its own consolidation. On American soil, on August 28, gushed the first drops of that precious liquid that would in large part determine the world's economic and political direction through the end of the twentieth century—petroleum. A year later the abolitionist Abraham Lincoln succeeded Buchanan. The United States then began its inexorable slide toward civil war.

Science was occasionally discussed in the salons. Determinism had not yet reached its peak or its most refined expression, since the universal principle of the conservation of energy, the discovery of which was completed in 1847, was still too new to be understood and discussed in larger circles. But already a new science, thermodynamics, was being discussed, encouraging new ways of thinking about transformation and irreversibility. In some well-informed salons people talked about the brand-new molecular hypothesis, stating that gases are formed from very small billiard balls in incessant motion.[2] In others they were surely talking about the mystery of those skulls and fossils found at Neanderthal that were thought to be almost but not entirely human.[3] Slowly but surely, except in the most conservative or fundamentalist circles, rationality, science, and progress were coming to be celebrated as a new holy trinity. In 1848 Renan proclaimed that "science is a religion; only science can resolve for man the eternal problems to which his nature imperiously demands solutions."[4]

Chance also played a concrete and visible role in daily life—not only in the games and lotteries that had always existed, but in other organized practices. The most striking example was in France, where, after 1818, the draft was the result of a sort of lottery. Drawing a bad number could mean six or seven years of service, of military campaigns and possible death on the field of honor. Money could ensure that this difficult destiny was reserved for the most unfortunate members of society through the practice of purchasing a replacement. Other practices also allowed misfortunes due to chance to be partially surmounted. By 1820, in effect, the divorce between the practices of insurance and gambling was finalized. The former became respectable to such an extent that life insurance became the favored example used when demonstrating the utility of probability theory. The insured person's wisdom and sense of sacrifice was held up as an example to those individuals who selfishly put their savings in a woolen stocking. Those who thought insurance might encourage employers to slight the search for optimal safety conditions still remained to be won over. For example, some reproached builders of steam engine boilers for preferring the security of good insurance to the steadfast research for technical conditions that would eliminate any accidental explosions. It was not until the end of the nineteenth century and the beginning of the twentieth that the respectability and even the desirability of insurance (on life, health, fire, traffic accidents, etc.) came to be accepted. More than ever, insurance companies could look forward to a prosperous future.

THE LEGACY OF PIERRE SIMON DE LAPLACE

Despite this new commerce with chance, the notion was at that time still viewed in a very classical way. Chance was seen as the mask of our ignorance, which science was charged with removing in all the areas it investigated. In 1859 the most celebrated and complete works on probability were still those of Laplace, dead for thirty-two years: the 1812 *Théorie analytique des probabilités* (Analytical theory of probability) and its lengthy but readable introduction, the *Essai philosophique sur les probabilités* (Philosophical essay on probability), written in 1814. They were far more popular than the more recent writings of Poisson, *Recherche sur la probabilité des jugements* (Research on the probabilities of sentences), written in 1835, and of Cournot, *Exposition de la théorie des chances et des probabilités* (Exposition of the theory of chance and probability), written in 1843.

In these essays Laplace, the author of *Exposition du système du monde* (Exposition of the world system), defended a purely subjective interpretation of probability, an interpretation linking it exclusively to our ignorance of the true states of the systems under examination. It could not be otherwise to such an ardent apostle of determinism. Remember his proud words on the omnipotence of calculus and reason: "Given for one instant an intelligence that could comprehend all the forces by which nature is animated and the respective situations of the beings who compose it—an intelligence sufficiently vast to submit these data to analysis—it would embrace in the same formula the movements of the greatest bodies of the universe and those of the lightest atom; for it, nothing would be uncertain and the future, as the past, would be present to its eyes."[5]

The "Platonic" attitude toward chance inherited from the century of enlightenment survived the introduction of probability into physics and biology, as evidenced by the contemporary works of Adolphe Quetelet. His research into probability led him to the happy discovery that the regularities predicted by the "physical" theory of probability were also to be found in social events—for example, in criminal statistics. From them he developed a study of a supposed "objective tendency to criminal behavior" that he claimed to be innate in each human and to be manifested only through chance. His detractors noted correctly that the observed regularities did not demonstrate the existence of such an ideal character. In truth, the variations observed from one sample of the population to another were too large to support this hypothesis, and they suggested instead that criminals are part of a special subpopulation, one that often relapses into crime. But these objections were not heard.

Almost the entire nineteenth century would be influenced by this theory of the "ideal man," individual representatives of the species being only more or less faithful copies. It took Claude Bernard to rouse the medical community forcefully against this ridiculous idealism and to point out that to understand humans, it is necessary to look not to the ideal average but to the real human, not to an abstract human model but to flesh-and-blood men and women, with all their individual variations. "There is no such thing as an average pulse, an average illness, an average urine," scoffed the author of the *Introduction à l'étude de la médecine expérimentale* (Introduction to the study of experimental medicine, 2:9). Bernard insisted on studying the actual biological characteristics of each individual. Is it not precisely by uncovering correlations between one and another variation that progress is made in understanding the functioning of the physiological system and in the practice of medicine?

Claude Bernard's biting irony rang the death knell for the Platonic conception of chance. Certainly chance has its laws, but those laws do not predate its manifestations. They arise from the manifestations themselves. Chance by definition allows no recourse to antecedents. Only events that are not determined by any cause can be called products of chance in the strictest sense. For that same reason, they constitute the natural limit of determinism, the upper limit of all the causal chains that can be reconstructed.

By 1859, however, the movement to surmount the difficulties of the classical approach to probability was already underway. In the years 1820–40, John Stuart Mill, Robert Leslie Ellis, and Jakob Friedrich Fries had elaborated the essence of the interpretation of probability in terms of frequency, an interpretation that still dominates the scene today. Shortly thereafter, in *The Logic of Chance* (1866), John Venn clearly expressed his doubts about the link between empirical regularities in problems of chance and the intrinsic formal causes of those regularities, such as the composition of the contents of an urn from which repeated drawings were made. But it was Augustin Cournot who made the first in-depth study of an independent chance freed from its Laplacian model.

THE WORKS OF ANTOINE AUGUSTIN COURNOT

The book of Antoine Augustin Cournot, a philosopher-mathematician from Burgundy, had as one of its aims to "make well understood the philosophical value of ideas of chance, of risk, of probability and the true sense in which the results of the calculations to which we are led by the development of the fundamental notions ought to be understood."[6] In fact, he was the first not to be satisfied with the classical interpretation of chance in terms of ignorance. In his work Cournot highlighted the most essential property needed when speaking about chance and probability, that of *independence*.[7] More precisely, he defined chance as *the crossing of two independent causal chains.* "It is a common sense principle," wrote Cournot, "that there are interconnected series of phenomena, dependent upon each other, and other series that develop in parallel or successively, without any dependence, any link of interconnectedness, with the first. Events brought about by the combination or the encounter of phenomena that belong to series that are independent in terms of causality are those that are called *fortuitous* events or results of *chance*."[8] Later Cournot would illustrate this idea by giving an example that is still famous today. "A tile falls from a roof whether or not I am passing in the street. There is no connection, no interconnectedness, no dependence between the causes that led to the fall of the tile and those that made me leave my home to mail a letter. The tile hits me on the head, and there is the old logician out of action. This is a fortuitous encounter or one that takes place by chance."[9]

Cournot's definition of chance is famous today, but it is nonetheless fragile. To a mind formed by Laplace's deterministic ideal, neither of the causal chains that determine the chance event through their fortuitous encounter can escape the universal determinism that rules the world. The independence of one of the chains in relationship to the other therefore can be no more than an illusion caused by an insufficiently global view. The encounter of these two chains thus can obey a superior causality and consequently could have been predicted.

Cournot did not overlook this objection. He responded in advance with a series of arguments arranged in increasingly general concentric circles.[10] On the one hand, he accepted that in some cases the postulated independence between the causal chains was not established. There the postulate was no more than the reflection of a pragmatic attitude. We might be able, he said, to see their common origin by retracing the two chains far enough. But in practice we can rarely do this. The calculations rapidly become too entangled to solve. This is why, rather than cry uncle, we invented chance as a category of explanation for appearances. In most cases the observed phenomena happen to accord with the laws drawn from this mental construction. That was the first circle. But, added Cournot, the cases of chance postulated through ignorance do not explain all the examples furnished by nature. The causal chains can also stem from distinct and unrelated physical laws. Cournot did not believe that the laws of nature can be assembled and summarized in one single formula, as later proposed by Einstein. He was convinced of the necessary and absolute independence of causal chains relative to different natural laws. That was the second circle. In this sense, he opposed Laplace's affirmation that an omniscient spirit could predict the fate of the entire world. This spirit would be superior only insofar as it could, much better than we, separate the *necessary* from the *contingent.*

In addition, and this was the third circle in his defense of contingency, Cournot thought that humans are capable of committing certain acts for no reason whatever and that this, in fact, is a general capacity characteristic of life. Thus, on the theoretical and epistemological level, he was led to propose the existence of phenomena contingent in themselves, maybe provoked by some physical cause but with no relationship of *reason* between them, because they are motivated by conscious life (e.g., deciding to go out to mail a letter). With him, then, we come to the extraordinary approach that leads us to infer, from the *absence* of a reason, the objective *existence* of fortuitous phenomena. According to Cournot, the absence of a reason is enough to validate the human invention of the concept of chance and the study of its laws. Cournot remarked that probability is actually a philosophical probability that "has to do with the exercise of that superior faculty by which we account for order and the reasons for things," because, faced with a given event, it is we who choose either to turn our attention to its chain of causes and effects or to assign it to the order of chance. It might be said that Cournot was, on the subject of "theoretical" chance, a true precursor of Darwin.[11]

Cournot's efforts were worthy, but they fell a bit short. In the end, they seem to lead only to an intersubjective notion of chance, valueless except within the framework of the universal relationships between thought and the world. Chance for Cournot remained an invention, not a discovery. Before being able to discover it, it would be necessary to reinvent it and then to use it in theories of nature. Maxwell and Darwin did just that. In the year 1859.

JAMES CLERK MAXWELL

Maxwell was only twenty-eight years old in 1859. He was not yet the respected and prestigious man of science that we picture today and that he became after the publication of the fundamental equations of electromagnetism in 1873. In 1859, however, without fully grasping his accomplishment, he made a discovery just as fundamental and just as important for the philosophy of science as that later achievement. Maxwell was, in fact, the first to associate a distribution of probability with the explanation of a physical phenomenon.

His interest in probability had been awakened a few years earlier by the physicist James David Forbes, his professor and friend, who discussed the existence of binary stars on a probabilistic basis. Maxwell then plunged into Laplace's writings (and Boole's) and mastered that new discipline. He also became aware of a long retrospective article written by the astronomer John Herschel and read Adolphe Quetelet's treatise entitled *Sur les probabilités appliquées aux sciences morales et politiques* (On Probability applied to the moral and political sciences [1846]). One passage from Herschel's article contained a demonstration of the least square method (which helps to determine the best straight line to be drawn on a graph through a cloud of experimental points, taking their uncertainties into account) applied to distributions with two random variables. This demonstration made explicit use of the idea that the two dimensions, *X* and *Y*, associated with the two random variables were independent—in other words, that their exact values and thus the fluctuations they displayed about their average values were not correlated. There, indeed, was a model that could be transposed to the motion of molecules. Their velocity or degree of agitation according to the three axes, *x*, *y*, and *z*, had to be independent. More precisely, it had to be stipulated that the projections of the velocities of the molecules according to these three axes each follow a Laplace-Gauss distribution, the most natural and general of the laws of probability (see chap. 3). Maxwell was able to deduce from this, for the first time, a correct determination of the energy distribution of molecules, henceforth called Maxwell's distribution. In so doing, he laid down the basis of the kinetic theory of gases.

In the race to discover the foundations of this new branch of science, Maxwell could have been overtaken by his elder of nine years, the German physicist Rudolf Clausius. In 1859 Clausius was a professor of mathematical physics at the Polytechnicum of Zurich. During this time he was working out the concept of entropy and its relationship to that of heat and temperature exchange. He had introduced the concept for the first time in an 1850 article, but he did not baptize it until 1865. In 1860 he engaged in a lively debate with Maxwell. The reasons for this dispute may appear to be of little consequence, but they perfectly illustrate the nature of the previous year's epistemological turning point. Pursuing his 1859 work, Maxwell established that the average relative velocity between two gas molecules is $\sqrt{2} = 1.414$ times the average velocity of each molecule, but Clausius insisted that he had demonstrated that the factor is only $4/3 = 1.333$. Looking more closely at their debate, it can be seen that Clausius's figure was derived using only the average velocities of molecules, whereas Maxwell had made the calculation complete by correctly taking into account the *probability distribution* of their velocities. Chance was understood once again to be an integral and necessary part of the equation.

Maxwell was the true founder of statistical physics. He introduced the idea of the statistical ensemble well before Gibbs, the American physicist who would systematize that approach. Curiously, the discoveries of this greatly talented researcher led him to a probabilistic and indeterministic philosophy, far from the materialistic ideal of Boltzmann, but common enough at the end of the nineteenth century. Maxwell's religious inclinations and his unshakable belief in humans' free will would finally make him a believer in dualism and vitalism.

Maxwell was also an essential link in the chain that extended the use of statistical methods in the social and life sciences toward their use in the physical sciences. As indicated by Lorenz Krüger,[12] while Laplace, Quetelet, and some others sought to found the social sciences on physics (if not to reduce the former to the latter),

Maxwell went in the opposite direction and laid the foundations of the new statistical physics of gases by making explicit reference to the works of the social statisticians. In particular, he recognized the qualitative link between the complexity of a phenomenon and the sources of variation that make the use of statistical methods pertinent as well as inevitable.

In 1859 Maxwell's other great challenger for the construction of the statistical mechanics of gases, Ludwig Boltzmann, was only fifteen years old. He had, of course, not yet developed his research program, which would later lead him to expand Maxwell's pioneering work by associating chance with irreversibility.

CHARLES DARWIN

On February 12, 1859, Charles Darwin celebrated his fiftieth birthday in his cottage at Down in the distant southeast suburb of London. Tired of voyages and worldly events, tormented by irregular health, he worked actively on the manuscript of his book *On the Origin of Species by Means of Natural Selection.* He had decided to draw conclusions from his extended studies, begun during his voyage around the world on the HMS *Beagle* from December 1831 to October 1836, as soon as possible. Rushed to set a publication date by the note Alfred Russell Wallace had just sent him, in 1858 Darwin exhibited some extracts of an 1844 manuscript and a letter he had written to the American botanist Asa Gray, dated September 5, 1857. These documents, followed by Wallace's communication, were read before the Linaean Society of London but created no stir. "The only public notice of them which I can remember," noted Darwin, "was by Professor Haughton of Dublin, whose verdict was that all that was new in them was false, and what was true was old."[13] Finally, in November 1859, his book appeared. The first printing of 1,250 copies was sold out the day of publication.

Unlike his first presentation, which was greeted with indifference, the book was quickly recognized as having great philosophical importance. Since then, the shock it aroused by suggesting the idea of a natural line of human descent, totally against the clerical dogmatism of the day, has often been noted. Remember the famous exclamation of a lady shocked to hear that "man descended from monkeys." "If this is true, let us hope that it does not become known," she cried. More significant from our point of view, Darwin's work was a new example of the use of chance as an element in explaining a phenomenon, and a vital phenomenon this time: the evolution of species.

Darwin was not actually the first biologist to use the concepts of chance and probability. Remember, for example, that a short time earlier (in 1856), Mendel had begun his systematic studies on heredity. Whereas Mendel saw probability as a complication made necessary by the play of hereditary combinations, however, Darwin was the first to take chance to be not an epiphenomenon but an essential constituent of the theory of living beings and of their evolution over the eons. For Mendel, chance was an encumbrance that obliged the experimenter to accumulate findings fastidiously. For Darwin, chance was the deus ex machina that, allied with natural selection, would reveal itself to be the driving force of evolution.

Darwin's true worth and perhaps his greatest achievement during his long deliberations on natural evolution lay in his having separated out, independently of the physicists, the irreducible aspect of chance—the superb, absolute indifference of the elementary events that underlie the appearance of chance events in relation to human judgments and the directives of destiny. As Charles Lenay put it, remarkably well,[14] the very hypothesis of chance introduced a rupture between two levels such that, each time randomness was invoked, it was henceforth necessary to specify that in relation to which the events were random. In the particular case of the evolution of species, Darwin underlined the independence between variation and the criteria of selection. He verified this idea first in the practices of breeders, in artificial selection, which can only channel the succession of obtained varieties, not cause or direct them. Then he extended it to natural selection. In this thought must be seen the first introduction of chance as an explanation in a biological theory. It is precisely for this reason that history, which is so often unjust, was right to name him, not Wallace, the founder of "Darwinism." Wallace had indeed identified the progressive evolution of species, but he had not analyzed its mechanisms as thoroughly.

On the other hand, although Maxwell had understood that local fluctuations, although improbable, can easily lead to observable heterogeneities, Darwin did not get interested in the effects of fluctuations on small populations, which would prove to be quite important in the creation of new species.

As to probabilities themselves, to chance itself, Darwin considered them to be an irritating obstacle in his pursuit of his personal vision. Did he not try, at the beginning of his scientific inquiry, to construct a description of biology and evolution as rigorous and mechanistic as Newton's description of celestial objects? This is why he speaks of probability only rarely and with an air of regret. When he does, it is to link it to our ignorance. He did not doubt that variations had unknown causes that would one day be unveiled. He did not fear that future enlightenment, as long as it did not contradict his fundamental axiom—the independence of the appearance of variations from the adaptation of living beings to their environments. For the time being, he regretted that ignorance of so many causes, especially when they gave birth to such marvels of exquisite complexity and harmony as the peacock's feathers. "The sight of the peacock's tail feathers, each time I see them, makes me ill . . . ," he wrote to Asa Gray in 1860.[15]

Darwin's epistemological program thus shows striking resemblances to that of Boltzmann. Both sought a mechanistic, Newtonian explanation for temporal irreversibility, the first in the phenomena of life and the second in those of the physics of gases. Boltzmann, like Darwin, found himself forced reluctantly to introduce probability—chance—into his works. For both men, the irreversible trend of the natural processes found its origin in the perfect independence of the elemental phenomena; variations from selection for Darwin, the elastic collisions of molecules from the global thermodynamic properties of gases for Boltzmann. We know that Boltzmann was one of Darwin's ardent proponents. It is not known, on the other hand, whether Darwin held reciprocal views of the illustrious Viennese professor. It is possible that he never paid him much attention. Darwin seems to have had little interest in mathematics. His notion of variation by chance was almost entirely uninfluenced by the formal development of the theory of probability. His effort of abstraction seems rather to have been of a philosophical nature. It could be said that by concentrating on variation in and of itself, by moving from an "essentialist" conception to a "populational" conception, Darwin cleared the way for the development, much later, of population genetics.

CHARLES SANDERS PEIRCE

In the wings of the theater when this play introducing chance was being performed, a young novice tried his voice. In 1859 he finished his studies at Harvard College. His father, Benjamin Peirce, professor of mathematics and physics at Harvard, encouraged him to pursue a career in the sciences, and Charles partially followed his counsel, although he was more naturally inclined to the study of philosophy.

In 1859, still a student, he was hired as an intern by the U.S. Coastal Survey (an agency that he would serve until 1891). The following year, however, he served another internship, this one on classification techniques under the direction of Louis Agassiz. Quickly he adopted a metaphysics of chance, considerably influenced by Darwin's book. His reflections on the role of chance in the physical and natural sciences pushed him to conceive of a *doctrine of necessity in a universe of chance,* which he delivered as articles in *The Monist* during the last decade of the nineteenth century. Although in this respect he was a great forerunner of the philosophical movement of the end of the millennium, Peirce was certainly too far ahead of his time. Despite the efforts of William James, no permanent university post was ever offered to him, and most of his writings were published only long after his death. Today well enough known in the New World, his work still deserves better recognition in Europe.

CHANCE IN PHYSICS AND BIOLOGY: A REVIEW OF THE SITUATION AT THE END OF THE NINETEENTH CENTURY

The construction of determinism by Galileo and Newton required that chance be relegated to the rank of an epiphenomenon, with no profound significance, that could be understood entirely by reference to the model of defective measurements. The fluctuations thus affecting measurements reflected above all the awkwardness of experimenters. As I have noted, this origin later dogged the notion's status during the entire nineteenth century. In sociology and medicine especially, this view led practitioners to the concept of the "average man," a concept whose dangers and excesses are obvious.

Maxwell and the other founders of the statistical mechanics of gases nevertheless did introduce a different point of view. This mechanics reduced thermodynamic concepts such as heat and temperature to molecular properties—heat to the total kinetic energy of the disorderly agitation of molecules and temperature to the average energy of each of those molecules. But this reduction has its limits. The absolute knowledge of the state of each gas molecule, its position and velocity, is not only impossible because of the imperfection of our measuring instruments; it is also inaccessible because of our practical inability to manipulate data in such quantity. At that time Maxwell apparently was ready to admit that this limitation contains something basic, something that gives to the principles of statistical thermodynamics a status different from what they would have if the introduction of probability in this theory had been tied only to chance rooted in ignorance.

The development of the applications of the theory of probability in biology caused a reawakening of sometimes radical questioning of the dogma of determinism for the thinkers who turned their attention that way. I have already mentioned the case of the physicist Maxwell, a profound indeterminist when it came to the phenomena of life and consciousness. Karl Pearson, the inventor of numerous statistical methods and tests that he applied to biology, emphasized that belief in strict determinism is no more than an act of faith. Charles Sanders Peirce was an even more ardent defender of indeterminism. His originality came in linking the idea of indeterminism to the property of temporal irreversibility more systematically even than Boltzmann did.

While the status of chance and probability evolved slowly, their fields of application expanded. In 150 years (1750–1900) the theory of probability invaded the domains of all the observational sciences to their furthermost frontiers. This invasion lent support to the idea that chance can exist even apart from our errors, that it can "be other than the name we give to our ignorance," according to Henri Poincaré. For reasons related to that awakening, a new philosophical way of thinking about human beings, stressing human contingency, arose and was constructed along a parallel track. Probability and existentialism seemed to walk hand in hand.

THE BIRTH OF PROBABILITY THEORY

3

The law of large numbers (or the laws of large numbers, since many variants exist) is undoubtedly the most important contribution of probability theory. It allows us to understand how chance, far from leading to anarchy, can produce its own laws by itself and without recourse to any hypothesis except for the essential property of the independence of chance events. For example, it explains why tossing a coin almost always leads to just about as many "tails" as "heads." However, these important findings of probability theory, completed in the second half of the nineteenth century by an entire arsenal of rules and mathematical theorems, left a certain number of more obscure but related problems unresolved. These problems, most often having to do with the manipulation of infinitely small probabilities (in some ways the theory's Achilles heel), clearly show that at that time the epistemological status of chance had not yet been entirely mastered.

THE LAW OF LARGE NUMBERS

The earliest and most general form of the law of large numbers is summarized in Siméon-Denis Poisson's communication before the French Academy of Sciences on November 16, 1835. "Things of every kind are subject to a universal law that we may call the law of large numbers. It consists in this: if one observes a very considerable number of random events of the same kind, depending on causes that vary irregularly, that is to say, without any systematic variation in one direction, then one finds that the ratios between the numbers of different possible outcomes are very nearly constant."[1]

Before Poisson, the stability of frequency in a series of repeated trials appeared to be a sign of divine intervention. Once the elaboration of the successive forms of the law of large numbers was begun, the need for special and external intervention was progressively eliminated. Frequency's convergence toward probability was then understandable as an inherent property of chance itself, implicitly written in the axiom of independence. This is in fact true, and it explains why it might be said that there is no chance without law.

Three stages can be distinguished in the progressive enunciation of the law of large numbers. Jakob Bernoulli's theorem (1713) indicated that in aleatory situations of a binary nature (tossing a coin, drawing balls from a vase containing only black and white balls, etc.), the probability of any difference between the frequency observed during a series of trials and the corresponding probability can be made as small as desired, provided that enough tests are run. In 1859, the very year Maxwell's works and Darwin's magnum opus were published, the Russian mathematician Pafnouti Chebychev generalized Bernoulli's theorem to apply to a series of tests on practically any random variable (as long as it allowed an arithmetic mean to be defined). Finally, in 1909, Émile Borel introduced the concept of almost certain convergence and demonstrated the law of large numbers in a very general form. In this form, the probability of a divergence between frequency and probability was reduced to zero for an indefinitely prolonged series of trials (a divergence can be produced only in a finite number of series where the probability of occurrence, in relationship to all possible series, is null). Appendix 1 gives some more technical details on these three successive forms of the law of large numbers.

From an epistemological viewpoint, the study of the various forms of the law of large numbers illustrates perfectly the inescapable role that the negligible event plays in probability theory, for it constitutes the fortress of the proponents of a subjective interpretation of probability.

On the whole, the objectivist view of the probability discussed here may seem the more politic, since it considers only a single operational concept, that of frequency. But this position by its nature remains incapable of offering a clear interpretation of the probability of a unique event (occurring only once).

CAN THE PROBABILITY OF A UNIQUE EVENT BE DETERMINED?

It is not always feasible to experiment on repeated series of events. In life also happy or unhappy events are encountered that have absolutely no chance of occurring more than once. I pose such a problem when I ask, for example, what is the probability that the Earth will collide with a large asteroid, similar to that which ended the reign of the dinosaurs 60 million years ago? Fortunately, experience can never let us verify the convergence of frequency with probability in such cases. Nonetheless, more general considerations of cosmology, similar enough to considerations of symmetry in other problems, warn us that an event of this type doubtless has a greater probability than we might wish. Astronomers recently tried to calculate the probability of such an encounter, taking into account the amount of debris from asteroids and comets that clutter the Earth's orbit. The figures they arrived at are, happily, very problematic. The fact that we are not certain of the future of the solar system, which is a complex system whose stability cannot be guaranteed over a scale of hundreds of millions of years, must also be taken into account.[2]

The probability of an event that has happened only a single time cannot be defined if the nature and the scale of all possible events is unknown. For example, on February 14, 1982, an American laboratory observed an abrupt jump in current through the coil of a superconducting solenoid constituting a "magnetic monopole" trap, the jump's value being compatible with the detection—the very first—of one of those still hypothetical particles in motion through the solenoid. Since no one has been able to repeat the observation, the possibility that it was an artifact cannot be ruled out, and the authors of the experiment have refused, for the time being, to insist on the validity of this fabulous discovery. Supposing that they had done so, the unique observation would not allow the probability of detection by their solenoid (the average number of detectable monopoles, per year, in this apparatus) to be determined. It would, however, have allowed an upper limit of such a probability to be established, which would already have been quite an accomplishment.

NEGLIGIBLE PROBABILITIES

The distribution of probability for a continuous random variable (which can in principle take on any value between minus infinity and plus infinity) that is most frequently encountered in nature is the "Laplace-Gauss" distribution, also called a "normal distribution," which stipulates that the probability that an event might differ from the average value assigned to it diminishes exponentially with the square of the divergence from that average. (This formulation is neither clear nor rigorous, but forgive me this weakness.) For example, normal law governs the dispersal of the results of many measurements of experimental studies when the sources of error are small and numerous and independently contribute to the total error. This finding was established by Laplace in 1810. The application of the normal law, not to the errors committed by a more or less able experimenter, but to nature itself, contributed to a reorientation of the interpretation of probability in an objectivist direction. I have already noted the success of Maxwell's hypothesis, which assumes that this law governs the distribution of the components of the velocities of the gas molecules of a given temperature on any axis of reference.

THÉORIE

ANALYTIQUE

DES PROBABILITÉS;

PAR M. LE MARQUIS DE LAPLACE,

Pair de France, Grand-Officier de la Légion-d'Honneur; l'un des quarante de l'Académie française; de l'Académie des Sciences; Membre du Bureau des Longitudes de France; des Sociétés royales de Londres et de Gottingue, des Académies des Sciences de Russie, de Danemarck, de Suède, de Prusse, des Pays-Bas, d'Italie, etc.

TROISIÈME ÉDITION,

REVUE ET AUGMENTÉE PAR L'AUTEUR.

PARIS,

Mme Ve COURCIER, Imprimeur-Libraire pour les Mathématiques, rue du Jardinet, n° 12.

1820.

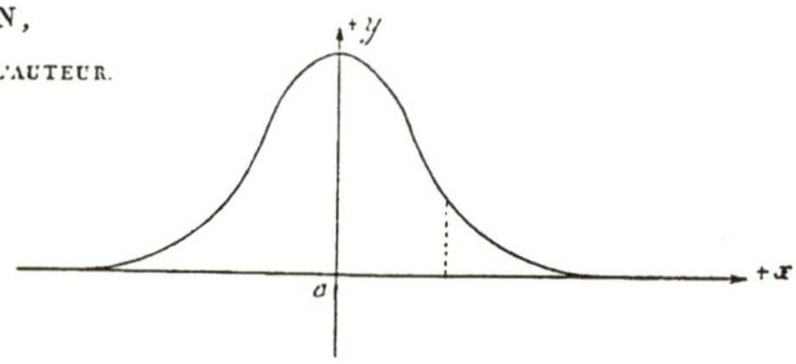

Fig. 1.

338 THÉORIE ANALYTIQUE

L étant une constante arbitraire; partant,

$$c^{-\frac{1}{2}\psi(x^2)} = c^{L-kx^2}.$$

Telle est donc la fonction qui peut seule donner généralement la règle des milieux arithmétiques. La constante L doit être déterminée de manière que l'intégrale $\int dx\, c^{L-kx^2}$, prise depuis $x = -\infty$ jusqu'à $x = \infty$, soit égale à l'unité; car il est certain que l'erreur x d'une observation doit tomber dans ces limites; on a donc

$$c^{L} = \sqrt{\frac{k}{\pi}};$$

par conséquent la probabilité de l'erreur x est $\sqrt{\frac{k}{\pi}}\, c^{-kx^2}$.

A la vérité, cette expression donne l'infini pour la limite des erreurs, ce qui n'est pas admissible; mais, vu la rapidité avec laquelle ce genre d'exponentielles diminue à mesure que x augmente, on peut prendre k assez grand pour qu'au delà de la limite admissible des erreurs leurs probabilités soient insensibles et puissent être supposées nulles.

Figure 1. Laplace's *Théorie analytique des probabilités* (Analytical theory of probability) explicitly demonstrates the law of errors. Here the author enunciates this law and is already concerned about the possibility of divergences tending toward the infinite relative to the mean. The drawing that illustrates the law is taken from an old treatise but does not occur in Laplace's book, which contains no drawings.

In a normal distribution the most probable value is also the average of the values the random variable can assume. The distribution around this maximal value presents a bell-shaped curve, with a bulge whose size depends on the concrete application symmetrically flanked on each side by a tail, which indicates that the wider a value's divergence is from the average, the rarer its occurrence, but that its occurrence is never completely impossible. The probability of the corresponding events, without ever being strictly nonexistent, can be as small as desired. Must we concern ourselves with these probabilities of events that are infinitely numerous but, taken individually, almost impossible? This is another difficulty with the theory. In 1930 Émile Borel believed that he could circumvent it by presenting what he called "the unique law of chance." This law declares that an event with an extremely low probability does not ever happen and that one can live or run experiments in the laboratory as if these possibilities were simply nonexistent.

Borel's attitude greatly simplifies life. For example, it is enough to suppose that a catastrophic event that is produced no more than once in a millennium on the average has no chance of affecting the same man twice in his life, which certainly makes for fewer sleepless nights. But is this attitude truly reasonable? How should one deal with events that have an infinitesimal probability of occurring but whose consequences would be incalculably immense? Imagine, once again, a large meteor striking the Earth's surface, plunging the planet into an extended glacial night. But think also of the calculation of probabilities affecting the safety of nuclear power stations, of which Émile Borel, who had the good fortune to live in an age without these enormous and complex mechanisms, was happily ignorant. According to Ivar Ekeland, Borel would probably have been wrong, as have most of the nuclear experts following him.[3]

In examples of this type, the pertinence of the classical approach, which insisted that decisions be made on the basis of the calculation of "mathematical expectation," that is, of average gain or loss, can be questioned. Certain events have no price or constitute an irreparable loss—death, for example. It is difficult in such cases to speak of average expectation or to evaluate an average cost. Psychologically, in any case, the criterion of mathematical expectation is often misleading. This was already known in d'Alembert's time. He opposed Daniel Bernoulli in 1760 in the case of "inoculation." The question was whether to support this practice, which foreshadowed vaccination against smallpox (but involved much higher risks than do today's vaccinations), based on statistics. D'Alembert reminded Bernoulli that many individuals prefer long-term risks, even if they are significant, to immediate risks, even if "small." The weakness in the classical calculation of probability is its superb disdain for the temporal dimension. It pretends to be unaware, here as in many other cases, that we give a greater significance to an event that happens today than we do to one that will happen in a more or less distant future.

THE SAINT PETERSBURG PARADOX

There is an instructive illustration of the difficulties that can arise when dealing with events of evanescent probability. Intrigued by one case of this type, Nikolaus Bernoulli submitted it by letter to Pierre de Montfort in 1713. The problem, called "The Saint Petersburg Paradox," has remained a classic in the theory of probability.

Let me translate it into modern terms. Suppose that Paul tells Pierre (who owes him a large sum of money) that he will forgive his debt if Pierre will flip for it. Suppose that Pierre chooses tails and Paul chooses heads. The usual rules of this game are modified. If the coin comes up tails on the first toss, Pierre has won and is totally free of his debt. If the coin comes up heads, however, he will have to pay Paul two francs, and the game will continue. Each time he loses, Pierre will have to give Paul double the sum given the last time. The game will be over as soon as tails finally emerges.

To figure out whether this proposal is to his advantage, Pierre calculates the expectation of Paul's gain. At the first throw, Paul has one chance in two of winning two francs, an average gain of one franc. He has only one chance in two that there will be a second throw and one chance in two that at that second throw he will win four francs, which is again an average gain of one franc. And so on. Each potential throw contributes one franc to Paul's chances for gain, and as the game could continue indefinitely, his mathematical expectation of gain is infinite. . . . According to this logic and contrary to common sense, no debt would seem to be worth playing by these rules!

But is it realistic to prolong the potential duration of the game infinitely? There are several reasons to doubt it.

First, Émile Borel's proposal stating that events of negligible probability do not ever occur offers a nonrigorous but certainly commonsense conclusion to the Saint Petersburg's Paradox: no game will continue long enough to ruin Pierre.

The second reason is more pragmatic and has to do with the nature of human savings, made up of coin of the realm. If Pierre loses the first n throws, he must give Paul a total sum of $2^{n+1} - 2$ francs.[4] He can allow himself to lose only a small number of times, or he will soon find himself insolvent. If he has 10 million francs, he could allow himself to lose only twenty-two consecutive times without going bankrupt. Paul's real expectation of gain, in the end, is thus only twenty-two francs . . . an altogether less attractive figure!

The last reason is the most radical. The Saint Petersburg problem shows how carefully mathematical expectation must be considered in some cases. The law of large numbers asserts that when mathematical expectation is finite, the relative difference between real gain and mathematical expectation tends toward zero as the number of plays increases. But in this case the mathematical expectation is infinite, and for every real play of finite length, the difference between the gain (necessarily finite) and the infinite mathematical expectation is always infinite. Thus there is no convergence in this case. It cannot be wise to be guided here by a mathematical law (the strong law of large numbers) when clearly that law does not apply to this type of case.

DOES THE DEFINITION OF PROBABILITY IMPLY A VICIOUS CIRCLE?

In his *Essai philosophique sur les probabilités* (Philosophical essay on probabilities), Laplace warned the reader that "the theory of chance consists in reducing all of the same type of events to a certain number of equally probable cases, or such that we are equally unsure of their existence," and defined the notion of independence in his third principle. "One of the most important points of probability theory, which lends the most to illusion, is the way in which probabilities are increased or decreased when they are combined. If the events are independent from each other, the probability of the existence of their combination is the product of their respective probabilities."[5] To take a concrete example, it is only by asserting that the results of successive throws of a die are independent from one another that the probability of throwing two sixes one after the other can be calculated to be $1/36$ ($1/6 \times 1/6$).

Thus, the axiom of independence is important because it is necessary to this essential expression of probability theory, that of the product of the probabilities relative to compound events. The most elementary statement of probability theory is impossible without the axiom of independence.

The mathematical theory of probability, as theorized by Andrei Kolmogorov, retained that stumbling block. The elementary events of a field of random events to which we want to assign probabilities must be independent. Let us return to the example of a player who throws a die. The appearance of a six or a two are two possible events, but they are independent in terms of their succession during a game. A good probabilistic theory of the game of dice could not be constructed if the appearance of a six affected the probability of the appearance of a two on the following throw (unless, of course, it would then be possible to pinpoint earlier, more elemental events falling under the hypothesis of independence of the different possible cases).

Like every fundamental notion of an intuitive nature, that of independence in probability theory is subtle and not easy to grasp. The tactic proposed in such cases is always the same. The property in question is supposed to be real and is given an operational definition. If the theory thus constructed proves to be an effective predictor, the status of the property advances from the theoretical to the ontological level. Thus, the property of the independence of elementary events ruled by chance was at first considered to be an expedient in the elaboration of a theoretical framework for treating games of chance, but it finally raised itself to the level of a fundamental principle of statistical physics (the *principle of equal "a priori" probability*), to such an extent that the statistician Richard von Mises was able to write: "Not even the smallest theorem of the kinetic theory of gases derives from classical physics alone. There is always, in addition, a hypothesis of statistical nature."[6]

■

Despite this difficulty (probability theory is not the only physical theory to face it; in fact, some epistemologists hold that every physical theory contains an element of circularity that is accommodated as long as the theory proves to be predictive), at the end of the nineteenth century, probability theory, already being applied to situations well beyond the simple, artificial games and situations that presided at its birth, encountered still another series of technical problems that were only poorly resolved. These were:

1. the likelihood of occurrence and treatment of evanescent probabilities;
2. more generally, probabilities having to do with continuous ensembles rather than discrete sets (the two problems are linked);
3. and finally, the practical interpretation of the results of the theory in a world dominated by fads, accelerated living, and the pertinence or obsolescence of values.

This intemporality of the principles of classical probability theory is not total. In fact, it tacitly concedes that the events whose possibilities it calculates are future events, and only the future can verify whether they occur. Other difficulties arise when one tries to apply it not to future events but to past events and when one evaluates the significance of those past events in terms of possible causes.

THE PROBABILITIES OF CAUSES

4

PREDICTION AND RETRODICTION

The man on the street would love to know whether the weather tomorrow will be nice. The roulette player would love to be able to predict where the ball might come to rest. The insurer offers a guarantee against misfortunes that might arise in the future. In everyday life, actions are considered based on what might happen, and our conscious thought is little concerned with past events. The past is no longer fearsome. It is significant that in ordinary language the word *prediction* is so openly oriented toward the future and that no word seems to exist for its opposite.

This is not, of course, because the past is of no concern; rather, it is because it functions differently. The past is the master, the source of all knowledge. Look at the way historians work. Most of the time they do not really look back to the past. Using documents, they try to bring life to a situation that has already been lived, and from that reconstructed moment they attempt to understand the succession of events that flowed from it. The naturalist method of establishing a chronology of the fossils found in geological layers is no different. In the West at least, this is the natural tendency of our minds: to learn from the past and to speculate about the future.

Another mental attitude is possible, however, one voluntarily turned toward the past. It consists of taking the present as given and asking questions about the possible (past) *causes* of what is displayed before our eyes today. Since there are many possible causes, the conclusions we reach are not generally unequivocal. Then it becomes necessary to weigh the various possible antecedents, to evaluate the respective probabilities. Using the neologism coined by Olivier Costa de Beauregard, we might say that we then face problems of retrodiction.

There is a fundamental asymmetry between the problems of prediction and those of retrodiction, one that must be grasped. In problems of prediction the laws of physics at least in principle allow a specification of exactly what the future state might be for any given initial situation. Only ignorance of certain details about the true initial situation obliges physicists to retain some level of uncertainty. In such a situation, they must settle for giving the probabilities of different possible final states, and the dispersion of the possible results will exactly reflect the lack of information with which they began. By contrast, a problem of retrodiction necessarily demands additional qualifications, beyond the intrinsic elements of the problem. These qualifications involve the relative probability that each of these causes was present before the event in question was produced. In other words, when the theory of probability is applied to problems of prediction, it requires only a limited view of the conditions that might lead to the realization of an aleatory event. The same theory applied to problems of retrodiction demands an extensive knowledge of the conditions that might have led to the realization of that event.

Bayes's theorem (see appendix 2) explicitly draws on these a priori probabilities, charged with summarizing the state of our general knowledge of the physical problem insofar as it might influence the final evaluation of the probabilities of causes. In other words, while problems of prediction might be solvable using physical laws, thus making a blank slate of all past life experience, problems of retrodiction require drawing on a sort of summary of all our acquired knowledge. To take a simple example, suppose that we find a red pebble on the ground and that we want to know whether it is a piece of bauxite or a piece of ferrite. Suppose also that our rather primitive methods of analysis allow us only to establish the probability that the pebble in question contains bauxite or ferrite. Would the absolute probability that this specimen is one or the other be the same if the pebble had been found near Marquette, Michigan, or near Little Rock, Arkansas? Of course not. The answer to the question would then depend on the a priori probability that a pebble, found where it was found, is of one type or the other, taking into account the region's geology.

Subjectivity thus enters Bayes's theorem, since in most applications there is no single way of determining the a priori probabilities. Thomas Bayes himself, for this reason, had doubts about the true import of his theorem, so much so that he did not publish it during his lifetime. His notes were published only by his estate executor, in a work entitled *Essay towards Solving a Problem in the Doctrine of Chance,* in 1763. Laplace made the same finding eleven years later, but unlike Bayes, Laplace was a disciple of the interpretation of probabilities in terms of ignorance, and the idea that ignorance could be translated into a uniform distribution of a priori probabilities did not trouble him at all. He published the theorem in his *Théorie analytique des probabilités* (Analytical theory of probabilities).

CHANCE AND IRREVERSIBILITY

It is in problems of retrodiction, or in the estimation of causes, that we can specifically experience the inescapable role of chance in temporal irreversibility. Just as Fourier's law on the diffusion of heat can be applied only to the future evolution of a system, the classical calculation of probabilities applies only when placed within the perspective of the flow of time and used for prediction. If these precautions are neglected, the calculations may sometimes result in striking absurdities. In *La Science et l'hypothèse* (Science and hypothesis), Poincaré illustrated this apparent paradox with an anecdote that I translate here in modern terms. If I sit down to a game of cards and see my adversary turn up a king on the very first card, what is the probability that this adversary is a Greek? (At that time the popular prejudice seems to have been that all Greeks were cheaters.) Classical calculation tells you that the probability is $^{13}/_{14}$ (since there is only one chance in fourteen that the first card drawn should be a king), but this is absurd, Poincaré argued, since if I thought I would run such a high risk of encountering a cheater, I would never have sat down to play.

The argument may seem less convincing in our day, when cardsharps no longer roam the streets. The physicist Satosi Watanabe gives a simpler example.[1] If anyone showed me a set of cards drawn sequentially in exact order from the deck (or in any other configuration arbitrarily specified a priori), I would not believe that this particular arrangement could have resulted from a blind shuffling, even if calculations insist on a finite nonnull probability for such an event.

Costa de Beauregard polished off this demonstration with particularly graphic examples:

> If anyone showed me a target pierced by a hole, I would conclude that a projectile passed through it, and never that a projectile will pass through the hole and restore the target intact. If I see footprints on the shore, I conclude that a man passed by there, never that a man will come, walking backward, and erasing the prints. . . .
>
> The procedures of retrodiction we have just considered are in no way blind. They are, on the contrary, among those always very delicate questions of the probability of causes, whose solutions do not just "leap to the eyes" without a very firm and very precise acquired knowledge, such as what is a bullet and what are footprints. . . .
>
> Footprints on the shore (or a bullet hole in the target) are witnesses of the original interaction from which the present situation arose, an interaction that must be specifically recognized in order for the problem of retrodiction to be resolved.[2]

Costa de Beauregard concludes that the a priori probabilities (which he calls Bayes's coefficients) are there to best represent the original interaction of the system under study with the rest of the cosmos, an interaction that, having ceased (practically) "at the initial instant," delivered this system into the state at which it is amenable to the calculation of probabilities.

It is well known that if one chooses among Bayes's coefficients as among equals, the formula for the retrodiction probabilities becomes identical to that of the probabilities of prediction. This is what Watanabe meant when he said that blind statistical retrodiction (which, by discarding our knowledge from the past, is often not wise) is temporally symmetrical with statistical prediction.

In summary, we can see that the classical calculation of probability in no way infers statistical irreversibility from premises that do not include it. Quite to the contrary, irreversibility is posited in the premises themselves, as an a priori asymmetry between prediction and retrodiction.

Costa de Beauregard ends with this caveat: "Since we have before us only a single planetary system, our own, the question of its origin is much further from being solved than that of the origin of the stars."

I have just raised the problem of the absence of a word in French or English to describe the effort of thought that, by somehow turning back on itself, questions the probability of causes. But is this really a rare activity for the human brain? Must we not, on the contrary, consider the brain to be a typically Bayesian instrument that ceaselessly evaluates the probabilities of causes?

Let us return to the practice of statistics for an example. It is clear that conscientiously updating mortality tables or the statistical records of other misfortunes would indicate the obsessiveness of a collector if they did not prove useful in other ways. In fact, their collection is a necessary prelude to the use of these statistics for purposes of reverse prediction, or retrodiction. The statistician accumulates these statistics to try to infer from them some new truth—for example, on the causes of death revealed by a heretofore unnoticed regularity. This is also the general goal of opinion polls and tests of hypotheses.

The idea of seriously considering the brain as an organ of statistical inference seems to have appeared around 1960, when it was put forth and developed by Ward Edwards in his work *Human Information Processing.* This book reports observations made on a number of subjects asked to give their opinions on the probable causes of diverse concrete situations (from a list of causes presented in advance). The experiment showed that in such situations, subjects generally adopt too conservative an attitude. They hate to revise their methods of judgment and do not give enough weight to experience. In addition, most of the subjects attribute too large a probability to the different possible hypotheses, so that the sum of the probabilities far surpasses 100 percent.

One of the first tests used can be described in the following way. Take two identical containers. The first, numbered 1, contains 70 percent red balls and 30 percent blue balls, and the other, numbered 2, contains the reverse proportion. You are presented with a group of twelve balls, eight red and four blue. What will you say is the probability that the batch was drawn from vase number 1 or from vase number 2? Naturally, Bayes's theorem can give you the answer by means of a bit of calculation, but that is not what is asked of you. We want you to reply "off the top of your head," to appreciate the competence of your brain as an estimator of causes. When averaged, the responses furnished by a real group of students indicated that there was a 75 percent probability that the batch was drawn from the first vase. The exact calculation, however, starting from equal a priori probabilities of 50 percent, gives a 96.7 percent chance that the batch in question was drawn from container 1. The disparity between the subjective estimation of 75 percent and the result of the exact calculation gives an example of that mental conservatism I just mentioned.

THE WEAKNESS OF THE SUBJECTIVE APPROACH

Remember that one of the first fields of application discussed for the calculation of probabilities was that of the credibility of eyewitness testimony in court. The following fictional situation has been used to illustrate the weakness of the subjective approach to probability theory in Bayesian situations.[3]

A taxi is involved in a hit-and-run accident. The town has only two taxi companies. The largest, which controls 85 percent of the taxis, uses green cars; the other uses blue cars. A witness testifies in court, saying that it was a blue car that fled the scene. The judge tests the witness's credibility by subjecting him to a series of reconstructions of the scene in conditions of comparable visibility. From a batch of taxis 50 percent blue and 50 percent green, the witness succeeds in correctly identifying the fugitive taxi in 80 percent of the cases. What is the probability that the taxi involved in the accident was in fact a blue taxi?

Ask yourself the same question and write your answer on a corner of a sheet of paper before reading further. The authors of the study showed that the students who were asked this question answered 80 percent on the average, and that is probably close to what you answered. But you can easily verify that Bayes's theorem gives only 41 percent chance that the taxi was truly blue if you take up the natural convention that the a priori probability as to whether it was a green taxi (or blue) was 85 percent (or 15 percent).

In reality, as with all Bayesian problems, this type of problem does not have a single answer, since the answer depends on the crucial question as to how you choose the a priori probabilities. The answer just given supposes that the absolute proportion of green or blue taxis in the city is taken for reference. But wouldn't it be appropriate to take the proportion of taxis of one or the other color that have had accidents in the past? Or that of the taxi drivers who have already been involved in hit-and-runs? Or the proportion of drivers of blue taxis who were in that area on that night? A different answer corresponds to each of these choices of a priori probabilities.

How do our brains really function in these problems of evaluation? It has been argued, with some psychological evidence, that the brain does not really weigh probabilities but rather clings sometimes to one certainty, sometimes to another, as in the story of the ruler who always agreed with the last counselor he had heard. Let us put ourselves, for example, in the mind of a woman asking herself anxiously whether she is pregnant. Is it possible to imagine that she would calmly calculate that she has a 70 percent chance of being pregnant? Wouldn't she instead go through a succession of phases of certainty that she is and certainty that she is not, spending perhaps over 70 percent of her time in the former state?

These considerations show us the danger and the limitations of any subjective approach to probability. It is probably true that probability is foreign territory to the psychology of the human brain, and especially when it comes to the probabilities of causes. But is our species not *Homo economicus*? Some have thought it might be possible to overcome the psychological difficulties I have just raised by anchoring the theory of probability in an economic approach. This approach gave rise to the notion of utility.

HYPOTHESES TESTING AND UTILITY

One possible way of overcoming the difficulties posed by Bayes's theorem is to refuse to be interested in the probabilities of causes in themselves but to focus only on the relationships between probabilities relative to this or that cause.

Thus, in 1928 the statisticians Jerzy Neyman and Egon Sharpe Pearson developed the theory of testing hypotheses. Subjectivity was then relegated to the choice of hypotheses to test and to the point at which the relative "utility" of the two hypotheses balance each other.

In 1935 Sir Ronald Fisher was even more dogmatic in proposing systematically to choose, as a counterhypothesis, what he called the "null" hypothesis—that is, the hypothesis that the phenomenon one proposes to display does not exist. Such an attitude, thought Fisher, would have to lead to a single response in any problem of inference.

In approaches based on the tests of hypotheses, one tries to minimize either an "error" (in the case of a scientific work) or an economic "cost."

Let us go back to the problem of the fleeing taxis. The witness could, for example, try to adopt an attitude that would minimize the risk of being wrong. This criterion would lead him eventually to testify in an excessively prudent, "conservative" manner, even if it was contrary to his first impression and to the evidence of his senses. For example, had he happened to know in advance the true proportion of green and blue taxis in the city (85 percent and 15 percent, respectively), he would testify in favor of a blue taxi only if this testimony was least likely to be in error.[4] An elementary calculation could then revise the true probability that he has correctly identified the fleeing vehicle. By taking this element into account, the probability that the taxi was actually blue is raised to 82 percent, a number close to the general estimations of the student "judges" tested.

The "utilitarian" approach to tests of hypotheses has in fact a more general scope consigned to "statistical decision theory," formalized notably by Abraham Wald. For example, it allows a level of confidence to be rationally defined. Given two hypotheses, the choice depends on the comparison of the costs of an error of judgment in one or the other direction. In this theory everything comes down to cost. For example, the engineer charged with receiving merchandise would establish the size of the sample to be tested and the scope of the analyses to which he intended to subject it in such a way as to balance the risk of a surcharge due to a faulty diagnosis of the quality of the merchandise with the costs of the analyses undertaken on the sample. Another example: why do you think that among physicists the established usage is to write, in the reports of an experiment measuring a size, only those decimal numbers the experiment can identify without ambiguity, plus the first uncertain decimal number? Because, answer the adherents of this utilitarian approach, one must stop at the point where the mention of a supplemental decimal can bring no benefit, justifying neither the price of the paper nor the reader's time.

Some authors seem to see in utility the beginnings of a possible reconciliation of the subjective and objective approaches to probability.[5] The true justification of probability would be to look in the fundamentally economic activities of its practitioners. Probability would thus be objectively . . . economical. I am not certain I can go this far with them. Chance existed on earth before the arrival of *Homo economicus.*

PART 2

CHANCE AND THE LIVING

VARIATION AND SELECTION: DARWIN'S NATURAL CHANCE

5

With the publication of *The Origin of Species* in 1859, the idea of fundamental chance unobtrusively entered the biological disciplines. Of course, it was not yet the ontological chance that quantum mechanics would later propose, but Darwin's work still marked an important stage. It broke with the idea that measurement fluctuations or errors necessarily belong to a multiplicity of causes and cannot be meaningful in themselves. To Darwin, the small differences or "variations" that appear from one generation to the other in the morphology (or more generally in the phenotype) of a species can indeed be produced by causes of which he was ignorant. The important point for him was that these natural variations are blind and that they are nevertheless the driving forces of evolution, since natural selection sorts out those variations that are advantageous. But this creation of direction depends on a rupture in the chain of causality between variations and selection.

BIOLOGISTS GRAPPLE WITH THE CONCEPT OF VARIATION

The earliest comments on variation in biology seem ludicrous when read today. In 1863 the German statistician Gustav Rümelin remarked that variability increases when climbing the ladder of life from the most simple to the most elaborate organisms. The learned, he noted in support of his thesis, are more diverse than illiterates, men more than women, Caucasians more than blacks, the moderns more than the ancients, and humans more than any other animal species. What must today's women and blacks think of such remarks?

In 1876 Wilhelm Lexis came up with the idea of studying social diversity mathematically. He wanted to analyze the dispersion of statistical data and observe its fluctuations from year to year. Annual data on the relationship between the number of births of boys and girls showed that it conforms to the predictions of the theory of probability: a fixed mean with average divergences proportional to the square root of the number of births. But Lexis also observed that when dealing with cultural data, the dispersion is always greater than that predicted by the theory. Such is the case, for example, with suicides, marriages, crimes, and births and deaths, all of which, to various degrees, follow the so-called laws of series, which is to say that they tend to appear in groups more often than do independent events.

At the beginning of the twentieth century, a systematic study of biological variation was undertaken within two concurrent schools of thought: the "biometric" on one hand and the "Mendelian" on the other. Although they both contributed to the development of statistical methods, and both were thoroughly anti-Darwinian in contesting the existence or the role of the causal rupture evoked by Darwin, these schools clashed bitterly.

In Europe the biometric school of thought was represented by Francis Galton, himself influenced by Quetelet. Quetelet had demonstrated that some natural variations—individual height, for example—follow the normal law of the distribution of errors, which had been created originally to describe the distribution of experimental errors committed during a repeated series of real measurements, such as the position of a star on the celestial sphere. In the minds of astronomers, there could be no doubt that the position measured had a true existence and a true value, independent of the observations. By analogy, noting that many biological parameters measured in a population are normally distributed, Galton believed himself justified in proposing that the observed natural variations also reflect imperfections, errors in relationship to a model whose existence is thus objectified. The "average man" is, to his mind, as real as the "true" position of a star is objective. For him, the biological sciences had to aim first at coming up with a description, as complete as possible, of that "ideal type," above and beyond the approximate embodiments of individuals.

The Mendelian school was just as anti-Darwinian. This school's approach rested, of course, on statistical reasoning and combinatorial analysis, but it placed at the heart of its reflections the idea that a causal description of variations had to be developed, not in terms of samples of populations, but for each specific individual.

It is interesting to note today the extent to which a discovery as essential as Mendel's (the combinatorial character of genes) was conceived not in an atmosphere of the crystalline purity of a cut diamond but by way of a fog of circumstance, of serendipity, if not even of peculiarities, among which the truth of the basic intuition would lead one to disregard such difficulties. In 1936 Ronald Fisher used the so-called chi-square test, invented by Karl Pearson in 1900, to determine the agreement or disagreement of genetic observations in relation to Mendelian predictions. In this way he was able to show that Mendel's original findings on smooth and wrinkled peas were in fact *too good* not to have been a bit "arranged." The proportions measured in each successive generation were too close to the relationships predicted by the theory for Mendel's experiment to be truly credible. The conclusion that Mendel's findings had been manipulated to seem more conclusive was inescapable. There was no question of casting doubt on the monk's personal honesty, but his too-zealous assistant was probably not entirely rigorous in deciding to which category to assign the different peas, letting himself, voluntarily or not, be influenced by a desire to arrive at the results for which his master hoped. Today we are well aware of the difficulty of eliminating biases of this type, all the more pernicious when they are unconscious. It brought about the institution, in medicine, of the "double-blind" technique for experimentation on drugs, where neither the subject nor the experimenter knows whether the substance administered is the drug whose medical properties are being investigated or a simple placebo. It can be argued that in Mendel's case, this bias only exaggerated a genuine phenomenon and that rectifying the findings would not have altered the conclusions to be drawn from the Austrian monk's experiments. That may be true, but how dangerous! In addition, we know today that the phenotype categories examined by Mendel (smooth and wrinkled peas) represent only a very exceptional case of genetic combination, since they depend in fact on a jumping gene. This affair will be studied in more detail in the following chapter.

As for Darwin, by 1859 he had already determined the meaning he gave to the term *variation.* We can sum up his thought in modern terms by saying that a mutation is *random* in that the chance of its appearance is not affected by its usefulness to the survival of the species. His reasoning could be further characterized in this way: I do not know whether chance in itself exists. But if variations are produced by any mechanism in such a way that they come about at random in relationship to their usefulness with respect to selection, then certain (observable) conclusions can be drawn from this fact. Thus, chance is defined by Darwin in relation to natural selection.

As we know, Darwin's thinking was predicated on a comparison between artificial selection, such as that practiced by farmers, and natural selection. According to most authors, the turning point in his thinking probably took place on September 28, 1838, during a period of intense theoretical research and following his reading of Malthus's *Essai sur les populations.* Malthus emphasized that the rate of population growth obeys a geometrical law of increase; in other words, in each generation the number of descendants is multiplied by a factor higher than one. Such populations overrun the territory at their disposal very quickly.[1] In most of the species encountered in nature, such exponential growth does not occur, and the number of individuals in fact remains just about constant. This observation means that there must be an enormous number of deaths in each generation.

What is the role of chance in variation? "We are profoundly ignorant of the cause of each variation," Darwin noted. "We are far too ignorant to speculate on the relative importance of the several known and unknown causes of variation; and I have made these remarks only to show that . . . we ought not to lay too much stress on our ignorance of the precise cause of the slight analogous differences between true species."[2] In a later work he expands on the same idea: "Throughout this chapter and elsewhere I have spoken of selection as the paramount power, yet its action absolutely depends on what we in our ignorance call spontaneous or accidental variability."[3]

Darwin's reflections on the role of chance were based on observations of the practice of artificial selection. What happens by chance in artificial selection is not only the variation in itself; it is much more the *utility* to be found in that variation—in other words, whether the variation serves the interests of the breeders. Chance's role in artificial selection is thus defined by those breeders. It might be said that in artificial selection the usefulness of the variation depends on chance.

Darwin illustrated his ideas by comparing the workings of natural selection to the work of an architect who builds a house by carefully choosing stones from those randomly scattered on the ground at the foot of a precipice. "The shape of each fragment may be accidental; yet the shape of each has been determined by the force of gravity, the nature of the rock, and the slope of the precipice—events and circumstances, all of which depend on natural laws; but there is no relation between the laws and the purpose for which each fragment is used by the builder,"[4] he observed. Whereas the architect makes his own choices, however, natural selection works mechanically and *only* once the variation has occurred. Thus, in nature, variation occurs by chance as judged from the point of view of selection, which is itself dependent on variations. This circularity poses a difficult problem in which is to be found all the subtlety of the notion of chance in Darwin's theory of evolution.

EPISTEMOLOGICAL CHANCE AND THEORETICAL CHANCE

In his thesis, Charles Lenay distinguished three successive levels of chance in nature: *epistemological* chance, *theoretical* chance, and finally *systemic* chance. Darwin did successively consider these three degrees of chance. The interpretation that he finally chose, and that appeared in *The Origin of Species*, was of the third type.

Epistemological chance is nothing more than the chance due to ignorance that we have already encountered. Thus, breeders do not know the causes of the phenotypical variations, but they do not doubt that these causes exist. Variations are thus random only from the breeder's point of view. The word *chance* here plays a role of substitution. It lets us avoid seeking an explanation. Lenay baptized a recognized ignorance "epistemological chance."

Darwin's theory laid down the principle that in wild species variations are at random *relative to natural selection.* To distinguish this new degree of chance, Lenay suggested calling it "theoretical chance." By definition, it is chance as judged from a theoretical point of view (the notion of natural selection) within the framework of a theory aimed at explaining a phenomenon (evolution). In fact, Darwin constructed this theoretical chance using the model of artificial selection's epistemological chance. When going from artificial selection to natural selection, he went from a use of chance in an epistemological framework to the use of chance in a theoretical framework. This clearly distinguished theoretical chance from epistemological chance.

But the role of chance in evolution did not stop there for Darwin. Central to his theory was the idea that variations become creative by orienting the destiny of evolution. Just as, in artificial selection, breeders continuously modify their criteria for selection in view of any fortuitous results from the preceding stage of variation, some natural variations make new selections possible.

The natural chance I am speaking of now is an active partner in the framework of the variation-selection system, since its effects can modify the criteria of selection. Lenay described this type of chance, a partner in a system, as "systemic chance."

CHANCE AND RIFT: THE CONTRIBUTION OF MOLECULAR BIOLOGY

The variations whose existence Darwin tried to prove had to be heritable to provide direction for evolution. "Any variation which is not inherited is unimportant to us," he wrote.[5] But he insisted on the existence of a time lag between variation and selection. The moments of variation and selection had to be distinct. Variation is produced by a cause's action on the reproductive systems of the parents. Selection is produced during the descendants' struggles for existence. Thus there exists a certain rift in causality, since natural selection works on the variations and not on their causes. Darwin saw this rift in causality as essentially characterized by this time lag. It could be argued that it was precisely on this point that his courage deserted him. True chance can be only a radical break and not just a distancing in time between links in the chain of causality.

The rift's function is to create the necessary independence between the variation's causes and the selection's criteria. How could a no longer epistemological but theoretical rupture be introduced in nature? Here Cournot's recipe might be applicable. The chance-brought event, the "utility of the variation," is found at the intersection of the variation's chain of causality, which has its own logic, and the selection's chain of causality. How can it be guaranteed that in nature there is no direct connection between the causes of variation and selection (the whole of the struggle for life)? By positing it. This Darwinian approach evokes Albert Einstein's fundamental epistemological approach, which has been called "theoretical realism."[6] Einstein thought that the ultimate reality of nature is hidden in the harmony of the equations, even though those equations are the fruits of human discovery. Thus, with Einstein, that which is (at first) freely posited may be identified with reality if it turns out to be in complete accord with nature. Darwin's approach was similar. The theoretical rift between mutation and selection, which is posited, is shown to be in complete accord with nature. For Darwin, there were two sorts of reasoning that explained this agreement. On the one hand, the temporal distance between mutation and selection contributes to making them independent, and on the other hand, selection depends not only on the individual but also on the external circumstances in which that individual is immersed.

It is important to recognize that this notion of a rift is distinct from—and in fact weaker than—that of independence, which is often evoked as a foundation for chance-brought phenomena. It is not necessary that there be absolute independence between the external environment of the living being that governs selection and the internal conditions that govern the appearance of variations. It is enough that this independence act in only one direction, a bit like a valve. The variations must not result in any way from selection, but the selection may result from, and in fact does often depend on, the variations.

This last comment, evoking the image of a one-way filter rather than that of an impermeable barrier between the levels of reality subject to chance on the one hand and to necessity on the other, between the chance of variations and the visible results of selection, between genotype and phenotype, as in physics between properties of atomic collisions and the sensible phenomena that are the objects of laws, is an essential breakthrough for human thought, for which credit must be given to Darwin in the field of biology. Without this breakthrough, we would never be able to assert anything that might characterize the more fundamental level, that of reality itself.

For Charles Sanders Peirce, Darwin's significance surpassed this already fundamental contribution. Peirce considered Darwin to be the precursor of the idea that chance creates order, along the lines of the concepts of modern epistemologists such as Ilya Prigogine, Henri Atlan, and Edgar Morin. "*The Origin of Species,*" Peirce wrote in 1893, "is responsible for the idea that chance begets order, which is one of the corner-stones of modern physics."[7] Peirce here alludes to the connection between, on the one hand, Quetelet's social physics and Maxwell's work in the statistical mechanics of gases and, on the other hand, Darwin's evolutionist thinking. Just as Maxwell understood the properties of the macroscopic order of gases in terms of the statistical distribution of individual molecular velocities, evolution must be understood in terms of variations and statistics on those variations. Alas, whereas Maxwell understood that local fluctuations, although improbable, could well lead to observable heterogeneities, Darwin failed to become interested in the effects of fluctuations on small populations, which would later be revealed as very important in the creation of new species, as we will soon see.

HEREDITY'S LOTTERY: CHANCE AND GENETICS

6

Our children resemble us in some ways, but they also differ from us quite significantly. Except with identical twins, they also differ from their brothers and sisters. In fact, all individual humans, despite the influence of heredity, display a unique mixture of the hundreds of thousands of genes received from each of their parents and reproduces that mixed genome[1] in every cell in their bodies. Chance plays a large role in presenting us with this certificate of originality.

We know that ordinary body cells divide through *mitosis.* In humans, for example, the DNA double helices of the twenty-three pairs of chromosomes present in a cell about to divide split along their lengths. While they are splitting, each half reconstitutes a complete chromosome through a pairing of complementary bases so as to endow each of the new generation of cells with a complete quota of twenty-three pairs of chromosomes identical to those of the parent cell.[2] In this type of reproduction, the new generation of cells is composed of copies that conform absolutely to their parent cells, except for rare errors made in the copying process.

An efficient way to introduce the essential variability between successive generations that the Darwinian model of evolution requires is for the new cells to have the characteristics not just of a single parent cell but of both parents. But the transmission of the parental characteristics to the children cannot be accomplished through a simple addition of genes. If that were the case, each child would have twice the number of chromosomes its parents had, and in a few generations the cells would be overrun by genetic material. The mixture of the parental chromosomes at the moment of fertilization must thus be preceded by an appropriate sorting mechanism, so that only half of each parent's DNA (only one chromosome from each pair) is present to be combined at fertilization. This is the system nature implemented with sexual reproduction. It depends on a special type of cellular division, specific to the precursor cells of eggs and spermatozoa: *meiosis.* Through meiosis the mixing of the parents' genes is effected with maximum efficiency, at random, as with the vigorous shuffling of a deck of cards.

Because the spermatozoa and eggs each have only a single set of chromosomes (rather than the two in normal cells), and because spermatozoa and eggs fuse and combine their genetic material at the moment of fertilization, the laws of heredity follow the rules of combinatorial analysis in the calculation of the number of possible combinations of the different elements, genes or chromosomes.

In 1866 Gregor Mendel, a monk at Brno in Czechoslovakia, published the results of his painstaking observations of seven genetic characteristics of peas. He demonstrated that the appearance of these characteristics in succeeding generations of peas crossed with each other followed the laws of combinatorial mathematics, if it was assumed that genes having the same function were present in two copies in each cell and that in case of conflict between the characteristics represented by these genes, some of them dominated others. He called the genes that dominated "dominant traits"; the others, which had to be present in two copies to manifest themselves, he called "recessive." Mendel sent his report to some of the greatest botanists of the age, but none of them gave credence to his work. Despite a continuing correspondence with the Moravian monk, the great botanist Nägeli did not find it necessary to refer to his works in the book on evolution and heredity he published in 1884.[3] It was not until 1900 that Mendel's laws were rediscovered independently by the German Karl Correns, the Austrian Erich von Tschermak, and especially the Dutchman Hugo de Vries. De Vries learned of Mendel's works only after he was already far advanced in his own research on the crossing of traits in several plant species and ready to propose laws equivalent to Mendel's.

Why did Mendel's work languish in obscurity for more than thirty years? Mendel, who was naturally humble, did not make much in public of what he himself called "an isolated experiment," even though he studied more than 30,000 plants over seven years! However, perhaps Mendel himself was troubled by the fact that the laws he discovered did not seem to be universally applicable. He was unable to relate them to all species and all characteristics. For example, Nägeli advised him to test his model by using the hawkweed *Hieracium,* which was unfortunate since that species reproduces also through parthenogenesis. In addition, the traits that Mendel chose in peas are carried by different chromosomes, so that he was able to avoid the complications caused by the possible cooperation of genes in the same chromosome and the crossing over of chromosomes (discussed later). A great deal of development was necessary before the Mendelian model of heredity could be generalized.

SEXUALITY INCREASES THE CHANCES OF VARIATIONS

Darwinian evolution requires numerous variations in a given population. Thanks to sexuality, the parental couple creates in its children what is for practical purposes an infinitely diverse number of variations based on the following theme: give our children half of my genes and half of yours. This way of transmitting heredity has the advantage of continually reshuffling a population's common genetic endowment in such a way as to ensure a certain homogeneity. The homogeneity thus preserved is, however, exactly the opposite of uniformity, because of the chance matching of each half of the mother's and father's gene pools.

Like the other body cells, spermatocytes and oocytes possess a complete double set of chromosomes. In these cells, however, during meiosis the pairs of chromosomes simply unpair, giving birth to two truly germinal cells, each with only twenty-three chromosomes: spermatozoa or eggs.

The unpairing of the pairs of chromosomes happens by chance. Since there are twenty-three pairs of chromosomes, the spermatozoa or the eggs made by a given individual can have 2^{23} different genetic formulas, or about 8.4 million. Knowing that a given woman produces 400 to 450 eggs throughout her fertile life, there is little chance that she will form eggs with even similar genetic coding. At the moment of fertilization, the spermatozoa's genetic endowment combines with the egg's genetic endowment. The number of different associations is then 2^{46}, about 70,000 billion, which is more than ten thousand times the total population of the world. This is the minimum number of possible variants for the genetic endowment of children of the same mother and father.

More precisely, meiosis is a sequence of two cell divisions accompanied by only one duplication of the chromosome material, which takes place at the beginning of what is called the prophase of the first division. Instead of leading to a simple doubling of the number of chromosomes, however, the two copies of the DNA remain glued to each other by a bridge, the centromere, making a single chromosome formed by two chromatids. In a species where ordinary cells contain *N* pairs of chromosomes, the nuclei generated by the first meiotic division of the spermatocytes and the oocytes thus each carry *N* chromosomes with two chromatids. Before the first division is completed and the double chromosomes migrate toward the two daughter cells, the analogous chromosomes (one originating from the individual's father and the other from the mother) superimpose themselves and somehow verify their similarity by congruence. This is very important in what follows. In superimposing themselves, they often are glued together at one or several points called *chiasmata.* During this pairing interchromosomic exchanges between portions of chromosomes (from one chiasma to another) frequently occur, and these exchanges, or *crossing-overs,* constitute a first and apparently totally unpredictable shuffling of the genetic material, even before the individual spermatozoa and oocytes are formed and combine in fertilization. After the separation of the chromosomes, the genetic material located between two chiasmata can find itself in either one of the separated chromosomes, like interchangeable building blocks. We do not know the nature of the physicochemical forces that allow this specific pairing.

The second division occurs without duplication. The two branches of each chromosome separate, each finding itself in one of the two nuclei produced by the division, in such a way that each could finally contain, through meiosis and its two successive divisions, four nuclei, each in its turn carrying N chromosomes with one chromatid, all already "variegated" by the crossing-over. This is effectively what happens for the spermatozoa. In females of higher animals and plants, however, one of the two nuclei created in the first division is expelled, leaving a "polar body" incapable of division. The same mechanism repeats itself at the second division. In the human female the meiosis of the oocytes leads to only a single fertile egg, carrying twenty-three well-variegated chromosomes.

The pairing that occurs during the first prophase of the first meiotic division seems essential to ensuring the success of sexual reproduction. In effect, it is because the pairing of their chromosomes is impossible that crossbreeding different species is unproductive even if it produces children in the first generation, as with crossbreeding donkeys and mares, which results in sterile mules.

Taking into account the secondary types of shuffling, such as point mutations in DNA, the transpositions of genes, and the jumps that some genes can take along the length of the chromosomes, along with the shuffling of the genes by crossing over during the first meiotic division and the sorting finally effected by meiosis between each pair of chromosomes, one can calculate that a single individual has the power to engender on the order of at least 10^{200} different spermatozoa or eggs. If we remember that our planet harbors no more than 10^{80} atoms, the number of possible variations vastly exceeds the number that can ever be realized. In this sense, the importance of chance in the concrete realization of a genetic endowment is obvious. If we add the chance social events that preside over the encounter of a given father and mother, we are, as are our children, truly the children of chance.

THE MAJOR TRAITS OF THE SPECIES ARE PRESERVED

Sexual reproduction is the elegant response nature found to deal with the need to multiply the number of variations, which are the basic materials of evolution, while still preserving the population's overall homogeneity. In a large population in which individuals have unlimited freedom to seek mates of the opposite

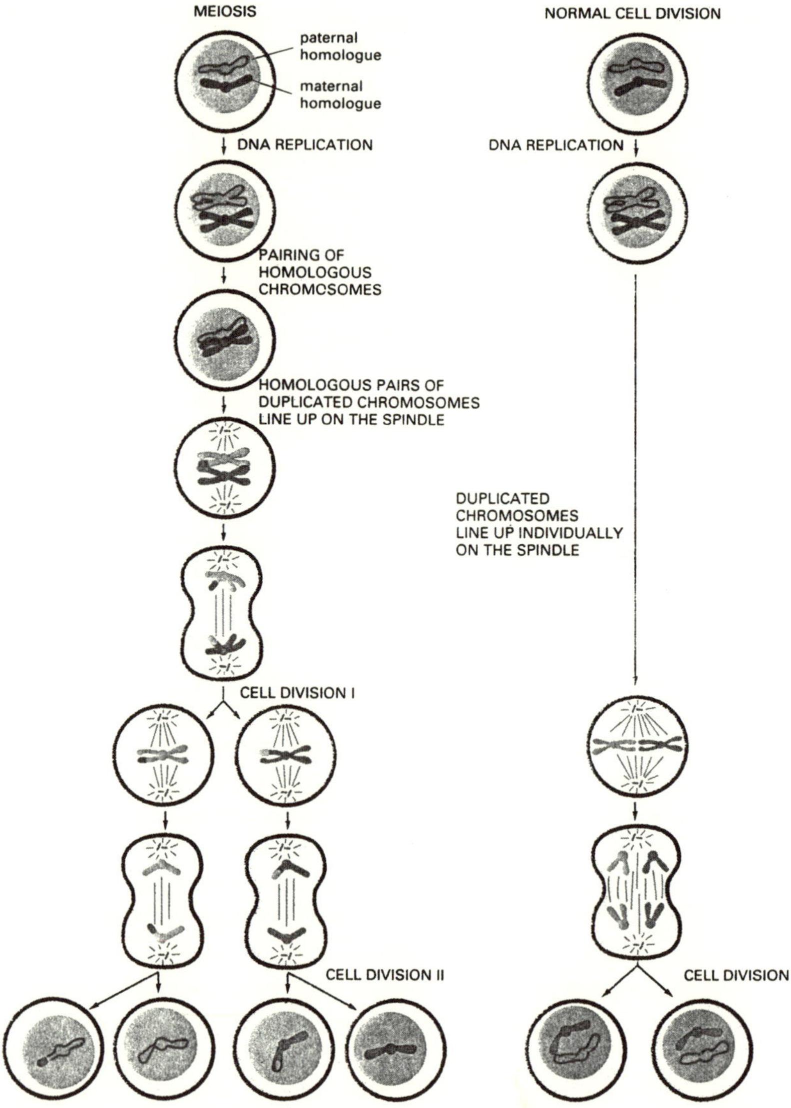

Figure 2. Comparison of the mechanisms of cellular division in meiosis and mitosis. To simplify the process, only a single pair of analogous chromosomes is shown. Generally, meiosis leads to four cells each with a single chromosome, each quite variegated. Mitosis reproduces the genetic endowment of the original cell identically in each of the two daughter cells. In humans the second division of meiosis happens only after fertilization, and the oocyte leads to only a single fertile egg, not four. (*Source:* B. Alberts, D. Bray, J. Lewis, M. Raff, K. Roberts, and J. Watson, *The Molecular Biology of the Cell,* 3d ed. [New York: Garland, 1994], 1015.)

sex among their own kind, this mechanism always results in homogenizing the population, with a bonus accorded to the most fertile genes, in accordance with Darwinian theory.

In practice, however, there exist many mechanisms that limit the mixtures produced by sexual reproduction. Geographical distance and the laziness of suitors can reduce the area used in the search for partners. Behavioral and psychological mechanisms that determine the choice of partners are other limiting factors, usually favoring couples with similar characteristics. Barriers of this type probably played a role, for example, in the appearance of the first hominids. As was suggested by the science journalist Marcel Blanc, we can imagine that the first humans—those naked apes—when they still formed only small colonies scattered throughout the large simian family, inspired some disgust in the apes still blessed with a generous amount of hair. The humans, for their part, were not ashamed of their nudity and were attracted to each other.[4] Later, the separation from simian ancestors accomplished, the human species expanded throughout the entire world and, through climatological and geographical accidents, was on many occasions isolated in distinct communities. These more or less complete isolations occasioned partial differentiation. Some groups, for example, acquired an elevated amount of melanin (black skins), while others produced much less melanin (white or yellow skins). One community geographically isolated by the phenomenon of glaciation during the Quaternary period nearly diverged enough to give rise to a new species. Many paleontologists think that a return to more clement conditions allowed that subspecies, the Neanderthals, to reintegrate before the divergence became truly consistent. The genetic pool of the Neanderthals was then lost in the larger pool of the Cro-Magnons, the invaders from the east. What a shame! Despite their receding foreheads, the Neanderthals had a cerebral capacity that was on average a bit superior to that of their conspecifics and even to that of today's human. A very recent study of the genomic endowment of Neanderthals tends, however, to conclude that species separation was completed long before Neanderthal extinction, 40,000 years ago. This is a most troubling result because one must then admit that two different species shared at that time an intelligence sufficiently developed to permit, for example, having elaborate burial rites.[5]

Today the human genus has, in principle, regained the ability to fully mix its genes, an ability that characterized the first nucleus of the human species. I remember a boy I met on a ship traveling around the world; he was searching for the face he'd seen in a dream and ready to go to the ends of the earth to find it. Through travel, cultural differences have a tendency to blur; all that is needed to measure the degree of homogeneity already achieved is to review the ethnographic reports of twenty years ago. The homogenization of the human species becomes possible once again. Will our distant descendants all be brown-skinned?

This is the panorama made possible by sexuality. By means of the ingenious mechanism of meiosis and its two successive divisions, nature ensures a continual mixing and remixing of genes. The first division of meiosis allows a preliminary redistribution of genes through crossing-over. The second division randomly separates the pairs of chromosomes that fertilization will later reassemble from the chance encounters between eggs and spermatozoa. But is that sexual dimorphism actually the only solution that nature could invent, or is it at least the best solution, since it is the one that natural selection finally chose? We are not sure, since over geological time some species have alternated phases of reproduction by way of parthenogenesis (with its strict preservation of genetic inheritance) and sexual reproduction. Aphids—those small soft-bodied plant lice—for example, use parthenogenesis during their invasion of a territory, sacrificing genetic diversity to population expansion. But this choice means a lesser adaptability for the species, a smaller chance of overcoming new environmental challenges.[6]

Sexual reproduction ensures the evolution of a species better than does parthenogenesis. As explained by Brooks and Wiley,[7] for a mutation to have a chance to be accepted in an asexual population, it must appear either simultaneously in a large number of individuals or at the moment when an external circumstance decimates the population that does not carry it. Sexual reproduction allows the rapid dissemination of the mutation. It also allows for maintaining a much larger biological compatibility between individuals, a greater *cohesion,* as indicated by the mechanism of crossing-over, which allows verifying, at each generation, that there is indeed compatibility between the genes coming from the father and the mother. Asexual populations would fragment more rapidly into subspecies and new divergent species.

CHILDREN OF CHANCE

We are thus truly the children of chance, and the lottery of heredity is there to organize this gigantic drawing of which we are the result.

Of course, not everything in our genetic endowment is the result of chance. There are also the constraints of heredity, the grouping of DNA into genes and chromosomes, which limit and delimit its effects. Each of us certainly has in our body several million molecules of the water that passed through Socrates' body (supposing that nature equitably recycles), and our bodies are made up of atoms, themselves recycled, that were forged some billions of years ago at the heart of a star. But we have resolutely preserved the genetic arrangements, the genes and the combinations of genes, that make for the characteristic traits of our families: blond or brown hair; Greek, Roman, or button noses; light or dark skin.

Overall, however, these preserved characteristics constitute only a small part of our diversity. One of the difficulties of the project to analyze the human genome is the extraordinary versatility of that genome. A very large number of genes possess alleles, or possible variations, and the identification of all these variations is tedious work.[8]

The common regrouping of these individual variations in larger groups (particularly into "races") does not have much biological significance. From the standpoint of individual variations, the "genetic distance" between two individuals chosen by chance from the same large human grouping is as great as that between two individuals taken from different racial groups. Let us not forget that racial differentiation is a fairly recent phenomenon, linked to the great migratory movements that took place during the last million years. In fact, it is generally thought that it does not go back more than two hundred thousand years, and may even disappear again if, with a little goodwill, we can surmount what remains of our disdain for cultural differences.

Humankind's mastery of space and the environment gives us this exceptional opportunity to avoid, for our descendants, a fragmentation of the species through the mechanism of species differentiation.

HOW ARE SPECIES BORN?

7

Despite the mixing of genes through sexual reproduction, groups of individuals often acquire genetic traits different enough from those of their species of origin that they break away and live apart. The rift is complete when this separation becomes so great that crossbreeding is biologically impossible. The separated group is then no longer a variety, a race, or a subspecies but is literally a new species.

There are many accidental causes of speciation. For example, separation can spring from external causes, either geographic or climatological, such as the formation of an isthmus or an island. But often the separation arises for behavioral reasons internal to the species. Individuals might suddenly change their taste in food or start looking for sexual partners with unusual morphological characteristics. A simple variation of a note in the love song of a bird or an insect may occasion the birth of a new species.

For Darwin, the direct descent of species did not count chance as an immediate cause. It was due, rather, to gradual changes imposed by natural selection. Today we are not so sure and are reexamining the exact role played by chance in the advent of new species.

The exact mechanisms of speciation, the biological "why" for the separation of species, are still poorly understood. As noted in the previous chapter, meiosis of oocytes and spermatocytes includes a phase of pairing of homologous genes (one originating from the father and the other from the mother) during which interchromosomal exchanges, or *crossing-over,* occurs. If the chromosomes from the two parents are too different, as is the case with the mule (born of a donkey and a mare), that pairing is impossible and blocks meiosis, making the hybrid sterile. But the details of the general mechanisms of speciation remain obscure. We can only suppose that certain mute parts of the genome, not translatable into proteins, play a role in barring interspecific fertilization.

Despite these uncertainties, we can still recognize the role played by chance in the appearance of successive lines.

DARWIN AND THE MODERNS

According to Darwin, speciation results from the continuous and obstinate action of natural selection. Variations are produced at random relative to selection, but selection itself owes nothing to chance; it reflects the state of the environment. This theory, still defended today by many biologists of the neo-Darwinist school, does not account well for certain striking aspects of evolution.

Today most biologists lean toward a nonuniform image of evolution, a succession of periods of stability interspersed with periods of rapid variation. These rapid variations may be triggered by different factors, such as the geographic isolation of small populations, the reorganization of entire segments of chromosomes, or a more delicate transformation of the genetic endowment. They could even be caused by a change in the interactions between the different loci—i.e., a modification in the permanent dialogue that takes place between the different genes and that often determines the appearance or nonappearance of certain traits not controlled by single genes.

Some species show obvious evolutionary changes, even over periods as short as a hundred years or less. We know about the appearance of species resistant to agricultural insecticides, the development of bacteria resistant to currently used medical antibiotics, and the adaptation of the colors on the wings of British moths to the blackening of the birch tree's bark caused by industrial pollution. On the other hand, some species show no or only minor modification of phenotype over a billion or more years. This is the case with some microscopic algae (blue algae) that are practically identical to the forms that populated the Precambrian seas. The brachiopod *Lingula* has remained the same over 450 million years. There are also some cases of "living fossils," such as the famous coelacanth, a fish that had been thought to have been extinct for 300 million years but was found living off Madagascar in 1930 and has since been found several times in the Indian Ocean.

Figure 3 compares a fossil dating from more than 200 million years ago with a coelacanth captured alive. Other examples include the mollusc *Neopilina,* which was thought to have been extinct since the Devonian era (330 to 380 million years ago) until several living specimens were pulled from the marine depths off Acapulco in 1957, and the conifer *Metasequoia,* fossils of which are currently being found that date from the Miocene era (25 to 10 million years ago) but which still lives in China. In 1984 an entire book was devoted to thirty-two studies of living fossils.[1]

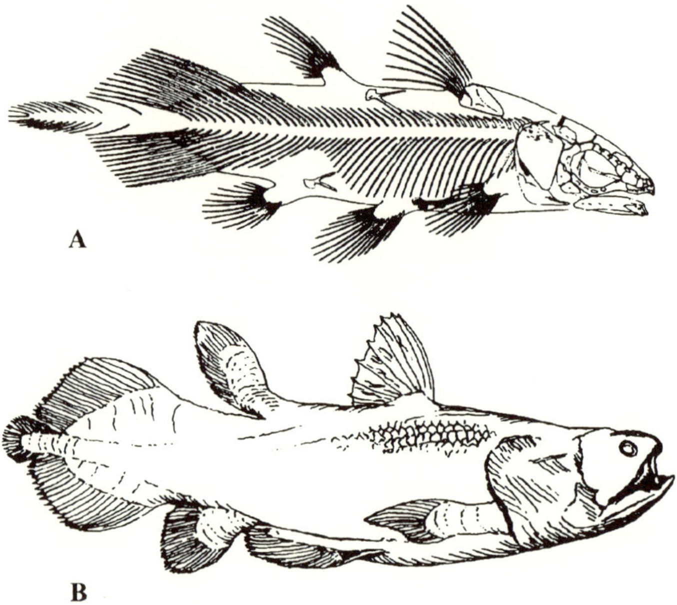

Figure 3. Comparison of the fossil coelacanth *Diplurus* (about 200 million years ago) and the living coelacanth *Latimeria chalumnae.* (*Source:* M. Kay and E. Colbert, *Stratigraphy and Life History* [New York: Wiley and Sons, 1965], 382.)

ONE HUNDRED SPECIES DIE OUT FOR EACH SPECIES THAT SURVIVES

The mechanisms of gene mixing linked to sexuality cannot in themselves be responsible for the abrupt accelerations of evolution required for the transformation of one species into another. Nonetheless, geological and paleontological findings show that over its history the Earth has nurtured about one hundred times more species than exist today. The enormous accumulation of fossils of extinct species in the geological strata has been known since Leonardo da Vinci's time (he contributed to their discovery) and caused a great deal of embarrassment to theologians once it became clear that humankind is no more than one of the latest-born on an Earth that had already witnessed the disappearance of many species well before our arrival.

The branching tree of biological evolution is often seen as representing a long, slow maturation of species in the warm oceans before life left that comfortable niche to conquer the surface of the land and diversify into plants, reptiles, birds, mammals, insects, and so on. It is sometimes easy to forget, as Stephen J. Gould insists, that the tree of life consists mostly of dead limbs and extinguished branches. Although it is true that today the Earth probably houses more species than ever before, most of those are only minor variants of a few large themes of anatomical organization.[2] The true image of the tree of life is not at all that of an umbrella pine, most of whose branches climb to the summit of the tree and spread out; rather, it resembles a leafless tree with only a few green, vigorous branches among a confusion of deadwood (see fig. 4).

It is estimated that there are currently about 1 million animal species on Earth. Insects alone represent about 700,000 species but are almost all grouped around a single archetype.

Figure 4. The tree of evolution is composed mainly of dead branches. (*Source:* S. J. Gould, *Wonderful Life* [New York: Norton, 1989].)

ISOLATION OF SMALL POPULATIONS AND GENETIC DRIFT

Just after World War II, researchers realized that the enormous differences in the speeds of observed evolution could be correlated with the population size of the species concerned (and possibly with the territorial range of their potential progenitors). For example, the emergence of the Isthmus of Panama caused almost no notable differentiation among the fish species living in the two oceans separated by that event, whereas minor geological events causing the geographical isolation of much more limited populations led to an obvious diversification of the animal species concerned. This is true of the bird colonies in the Pacific islands that so captured Charles Darwin's imagination. In 1836, using the collection reported by Darwin after his tour of the world, the ornithologist John Gould distinguished no fewer than thirteen different species among the finches of the Galapagos Islands. They differed mainly in beak size and form. Darwin was easily able to show that these traits were related to the habitat of each colony and to the type of food that was found there: harder or softer seeds, types of insects, and so on. It became obvious that these finch colonies were all derived from a common species that adapted itself progressively to the conditions prevalent on each island.

The most important aspect of population size is that it affects the selective value of certain genes. Thus, in a small isolated population, homozygosity (the fact that the two homologous genes carried by a pair of chromosomes—one originating from the father and the other from the mother—carry the same trait) can play a much more important role than it will in a large colony because of the higher rate of consanguinity manifested during breeding. In contrast, once a population has reached a certain size, it generally experiences a higher rate of sexual mingling and seems to be much less apt to produce new species where significant new evolutionary traits appear. The much more efficient genetic drift of small populations can also cause the total elimination of genes and the impoverishment of the species's gene pool.

Speciation is a frequent phenomenon, and the lifespan of the majority of species is very short. It is only in rare cases that the new species possesses a net advantage in selection over the species which gave it birth. In such cases, it tends to displace the parent species in the ecological niche the latter occupied.

With most animals the sterility barrier is not the immediate cause of species isolation. In many cases behavioral isolation has already been established long before the infertility barrier appears. Of course, from a taxonomical point of view, infertility is the criterion that is used to characterize a completed differentiation. From biology's point of view, however, ethological barriers are more important. For example, the signals to which sexual partners must respond are coded in the genetic program of each species. A mutation in the gene coding the rhythm of the male's song may be enough to ensure that those affected will pair off only with a limited female subpopulation, thus favoring the mechanisms of genetic drift and speciation.

The role of behavioral adaptations is particularly emphasized by Ernst Mayr.[3] Male animals make their displays before females of their own species, and conversely females typically respond only to the display of males of their own species. If, for example, a just-hatched female drosophila is kept in total isolation until she is ready for reproduction and then given a simultaneous choice of males of several species, including her own, all those males display before her. But almost without exception, she accepts the courtship only of the male of her own species. In fact, in most cases several characteristic signals must be present simultaneously to bring about recognition of the chosen mate.

Mayr gives two striking examples of behavioral Darwinism. "In the turkey cocks, the chicks have to utter a characteristic call. If they fail to do so, they are killed by the mother hen. If one deafens the hen, she kills her own young. . . . In many other species, and particularly in the domestic fowl, there is no instantaneous recognition of one's own young. Everyone is familiar with the picture of the mother hen leading a brood of little ducklings." This is the phenomenon of *imprinting*, which Konrad Lorenz meticulously observed.

In large bird colonies and troops of mammals, the ability to recognize the species is not enough. To care for their young, the parents must be able to recognize them among a large population of young. From the neo-Darwinist point of view, this capacity for specific recognition is a selective advantage, since it favors the transmission of a genetic endowment closest to that of the parents. The reverse is also true. The children have an interest in recognizing their parents as soon as possible from among the multitude of similar adults. Such individual recognition exists shortly after birth among mammals and two to five days after hatching among birds living in large colonies. Some young are even capable of learning to identify their parent's calls before hatching. In large colonies of seabirds and Antarctic penguins, the precision with which parents can find their children is almost stupefying. A series of films by the Cousteau Foundation documents these marvelous mechanisms of recognition.

The parent's faculty for discerning and memorizing a youngster's individual characteristics from the moment of its birth or an offspring's ability to recognize its parents requires that the species be endowed with an elaborate genetic program while remaining flexible enough to adjust to individual variations. As Mayr pointed out, such a complex program could not be rigidly inscribed in the genetic endowment, since the interplay of genetic recombination would destroy it with each generation. It must be learned *after conception.*

It is in this direction that cognitive psychology's most recent findings on nurselings are leading, according to a recent report by Jacques Melher and Emmanuel Dupoux.[4] They say that young humans are highly receptive to the most diverse sounds, so that even before birth, they have an astonishing capacity to distinguish random noises from the language used by their mothers and to recognize objects by their form. In short, the young human has the ability very rapidly to evolve a grammar, even if summary, and a science of the meaning of symbols (a semiology) and their combinations, whether those signs are aural, visual, or olfactory.

THE CEREBRAL EXPLOSION IN HUMANKIND

Once again, evolution considered in its entirety and on the scale of species has not been regular. It seems rather to have undergone long periods of stagnation, punctuated by brief flashes during which it was incredibly accelerated. For example, bats appeared from a type of insectivorous mammal over the course of a few million years but have undergone no structural modification in the 50 million years since. The changeover from the thecodont reptiles that populated the Earth 225 million years ago to the archeopteryx with feathers and the ability to fly took only a few million years, but modern birds as a class have changed little since their appearance about 70 million years ago.

These periods of accelerated evolution, during which new structures appear as if thrust forward by a strong wind, raise some formidable questions for the Darwinian evolutionist. How can natural selection explain, for example, the appearance of wings in birds, when the precursor organs don't seem to have any selective value before they are sufficiently developed to be functional? According to Mayr, this would be possible only if the embryonic organs possessed, for a time, a double function, the new function finally outweighing the old. One possible example of such an evolutionary path is furnished by the gill-lung organ pair. Some intermediary forms of those organs could have been very useful in the struggle for survival in certain prehistoric environments.

But the most striking example, and the most important to us, is incontestably the fast expansion of human cerebral capacity during the last million years. There are some indications that this evolutionary stage might have ended about 100,000 years ago, having endowed the human genus with the largest brain in animal history, with the exception of elephants and dolphins. The cessation of human evolution, if it is real, could have two distinct causes. One of them might be that humankind has become unaffected by changes in the environment, insofar as we have mastered them culturally rather than submitted to them passively. We clothe or unclothe ourselves, warm ourselves or create air conditioners, and raise grain and create new plant or animal species as a function of our needs and the changing requirements of the environment. The supposed prime mover of natural selection, the survival of the fittest, thus has lost its power. A second reason could be that humans have (recently) developed transportation so much that the mechanisms of geographic or genetic isolation that spur the appearance of evolutionary niches are disappearing.

The presence of periods of accelerated evolution inspired François Jacob's and Yves Coppens's beautiful reflections on the theme of *opportunistic chance,* and it is indeed thus that chance must be seen to work in nature, especially when it deals with life. Chance is, perhaps, not directed toward finality or a teleonomic project, but it "knows" how to profit from the occasion—from all the occasions that present themselves—to work toward the development of life, the complexification of the organism, the expansion of intelligence, and the increasing autonomy of the living system. Gaining autonomy: according to some, this last point is the true and only *goal* that can be imputed to life.

GENETIC GRAMMAR AND VARIATIONS

Molecular biologists are often fervent defenders of neo-Darwinism, while paleontologists are more reserved. Those who have acquired competence in the two domains propose substantial modifications to orthodox theory, especially when it comes to the alternate phases of stagnation and accelerated evolution (Kimura's neutralist theory and Gould's punctuated equilibrium). The novelty arises from the paleontologist's greater consideration for historical constraints that harness chance and build, little by little, a semantics or a grammar of evolution, giving it direction. By carrying this logic to its extreme, one can, along with Brooks and Wiley, turn the notion of evolution around to understand it, paradoxically, as life's time lag relative to the burgeoning expansion of the phase space[5] available to living species.[6] In fact, as quickly as life occupies new niches and living organisms become heavier, they reorganize more and more matter. In becoming larger, organisms are not content just to accumulate matter; they arrange it in cellular units and then increase the number of cells they incorporate in the cybernetic and regulated networks of the organism. Each of these stages contributes to multiplying the available phase space by a large factor, a process much faster than a simple aggregation of unformed matter. In view of constraints due to the modes of organization of life and of those linked to heredity, which channels those forms of organization from generation to generation, living matter does not in fact occupy more than a small portion of the physical phase space available to it. In fact, the relative size of this portion is continually decreasing, as quickly as the organism becomes more complex. This, argue Brooks and Wiley, is the source of the illusion that life defies the laws of thermodynamics, for it creates systems that are always farther from a state of equilibrium relative to an explosively expanding phase space.

The grammar I am discussing is not the result of historical luck. It can seem that over the course of millennia, life has become charged with an immense memory that henceforth will channel all possible innovations. This memory, aided by the patience of centuries, would reflect the ineffable qualities of the environment. Thus, life would have been, bit by bit, compelled to evolve in a direction that was not arbitrary but that would already have been inscribed on the heart of nature, in the subtle interactions tied ultimately to heat exchanges, since these are the ultimate sources of irreversibility.

If this vision is true, we would expect to see life try many directions, hesitate, take off and return, and then set out finally in one direction and accelerate progressively. This is indeed the way things have happened. It took one or two billion years to pass from infusoria to the multicellular stage. Some hundreds of millions of years sufficed for the passage from microalgae to the first vertebrates. Mammals and birds took only some tens of millions of years to appear, and humans, less than ten. Our view of history is too myopic to know whether evolution will continue despite perturbations caused by our actions affecting nature, or whether, with humans, evolution has found its final expression.

Darwin launched a school of thought destined for greatness but that he himself was not able to exploit to its limits. In the first decades of this century, E. Mayr advanced the idea that in our studies of evolution, we have to take into account not only individual genetic variations but also the conditions under which these genetic variations manifest themselves. These conditions include the size of a population, its relations with neighboring populations with which it is in competition for natural resources, customs regulating behavior in each society, and so on, all of which play a role. The biology and the genetics of populations were thus able to develop into autonomous disciplines, essential to understanding the history of life and the challenges that confront it today.

■

At the end of this review of the implications of evolution and the mechanisms of speciation, biological chance looks very different from the chance for which gamblers and mathematicians had prepared us. It is not the passive and indifferent chance of games of dice or of Kolmogorov's interminable series of incoherent signals. Not that natural chance, considered in itself, can remain untouched by the properties of versatility and absence of memory that define mathematical chance; those properties must necessarily form the essential ingredients of a true chance. Nonetheless, incarnated in matter and enrolled by life in the historical process of biological evolution, natural chance is revealed to be endowed with its own infinitely subtle qualities, which might be called emergent qualities. These are the qualities that François Jacob, in *La Logique du vivant,* proposed summarizing with the word *opportunism.* Natural chance is opportunistic. In other words, it is attentive to its own unfolding, so as always to offer life the opportunity for greater expansion.

IMMUNOLOGICAL DARWINISM; OR, HOW ARE ANTIBODIES BORN?

8

Evolutionary chance requires first and foremost that we think about life in terms of "populations" in all their essential diversity. Mutations, Darwinian variations, and the gene mixing achieved by sexuality all ensure that we live in a world that is not, perhaps, the only possible world but that is certainly very different from what it would be without these aleatory mechanisms. What catastrophes, population crashes, and wildfire plagues would devastate a world inhabited by genetically identical colonies of animals? Each of the aleatory mechanisms I have already discussed allows chance the opportunity to manifest itself in the living world. Examples of chance's workings are also to be found by observing the immune and central nervous systems.

The immune system is the ensemble of genetically based mechanisms with which the body defends itself from outside attacks. It furnishes two examples of aleatory mechanisms: the fabrication of antibodies and the recognition of self versus nonself. This latter function is ensured by a particular gene system: the major histocompatibility complex. This system's ability to produce a large variety of molecules is made possible by the incredible versatility of the corresponding gene system, which is an especially striking case of genetic variation (having a great many alleles, alternative forms that occupy the same chromosomal locus).

The variety that appears in antibody fabrication does not reflect a similarly large diversity of genes. In this instance nature uses a small number of genes that do not necessarily show alleles; rather, it makes use of the enormous variability of their combinations. The proteins that are called "antibodies" are assembled from small fragments spliced together like colored beads on a chain or mosaics of vivid and contrasting colors. Observing their fabrication gives us another opportunity to witness the proficiency of chance at work in the living world, constructing its products from odds and ends but also remembering its successes so that each subsequent construction can build on experience. Let us now turn our attention to this process of fabrication.

Immunology has only rather recently (in the past thirty years) elucidated the mechanisms of antibody fabrication from a limited gene pool, which follows a succession of trials and errors among randomly assembled elements that multiplies the probabilities of success in combat against the alien enemy. Briefly, antibodies are built through the random assembly of elemental building blocks. This haphazard construction allows the elaboration of numerous molecules in forms adapted to latching onto alien antigens and marking them. Only those molecules that prove effective in the fight against these antigens are then reproduced.

A FORMIDABLE MEANS OF ADAPTING TO AN AGGRESSOR

The body is constantly subjected to various forms of aggression. Bacteria can penetrate and multiply, producing toxins that are devastating to the organism's normal cells. Viruses can insinuate themselves into the cells, parasitizing them and forcing them to express the virus's genes. These bacteria and viruses carry characteristic molecules on their surfaces: antigens. The immune system uses those antigens to recognize aggressors and determine which antibodies to produce to combat them.

The molecules of ordinary proteins that are fabricated by our bodies from a given gene are all identical. They have exactly the same sequence of amino acids. In contrast, the antibody molecules produced by the immune system may have millions, even billions of formulas that differ to some extent one from another.

Lymphocytes (white blood cells) in the blood are the principal agents of the immune response. They can be separated into two groups: B and T lymphocytes (or B cells and T cells). Whereas the first group recognizes the antigens circulating in the blood, the second recognizes and fights the body's cells that are already infected.

A pathological bacterium or a virus generally carries, on its surface, a large number of antigens that the B lymphocytes can recognize. There exist several types of B lymphocytes, however, each with its own function. Some specialize in the direct production of antibodies. Others specialize in memory and the recognition of past aggressors to ensure a faster and more intense response in case of recidivism. Finally, a third group, called "regulator B lymphocytes," functions as either an amplifier or a suppressor of the immune response.

Antibodies fabricated by the B lymphocytes stay stuck to the surface of the lymphocyte's external membrane. When an antigen links with such an antibody, the lymphocyte is stimulated and begins to proliferate (by multiplying itself through successive divisions).

The T lymphocytes recognize the cells marked by the B lymphocytes and attack them. Some T lymphocytes play the role of regulators, allowing acceleration or deceleration of antibody production. Others, the effector T lymphocytes, destroy the aggressors—eliminating tumorous cells, for example—or kill the cells infected by a virus. For this reason these lymphocytes are called "cytotoxic" or "killer" lymphocytes. Finally, there is a category of T lymphocytes responsible for the reaction called "delayed hypersensitivity." These lymphocytes react with a variable delay, up to twenty-four or forty-eight hours after the invasion, and call for a maximal response, for example, only on the fourth day. They warn of new outbreaks of infection and are charged with getting rid of any hidden sources of recurrence.

SPECIALIZATION OF LYMPHOCYTE POPULATIONS

Let us imagine a B lymphocyte surrounded by antigens. An antigen sticks to its surface, and the lymphocyte begins to produce antibodies. It can also be stimulated indirectly by a T lymphocyte that has recognized the antigen in an infected cell. In such a cell the infection is expressed not only by the presence of antigens at its heart but also by changes in its external membrane, which carries the characteristic antigens of the infection on its surface. In these circumstances the T lymphocyte activates a B lymphocyte that in turn produces antibodies. Some B lymphocytes transform themselves and become memory lymphocytes, preserving the record of the pertinent antibody-antigen association.

Antigens come in many varied forms, and each form calls for a specific response. This response consists of the production of an adapted antibody, typically an immunoglobulin.

An immunoglobulin is a large Y-shaped macromolecule (fig. 5) formed by four polypeptide chains, two heavy and two light. Disulphur bridges link the heavy chains to the light chains and the heavy chains to each other. The heavy chains consist of a constant zone, the same or almost the same for all antibodies, followed by a zone that varies according to the type of antibodies to be synthesized. These constant zones form the base of the trunk of the Y-shaped immunoglobulin, whereas the variable zones form the tips of the molecule's branches. Similarly, the light chains have both a constant and a variable zone arranged in the same way on the two branches of the Y. The antigen reception sites are located at the ends of the immunoglobulin's Y branches.

Let me be more specific. The heavy chain's constant zone contains about 330 amino acids; its variable zone, about 110. The variable zone includes hypervariable zones that differ from one molecule to another. The constant part is characteristic of a certain type of immunoglobulin and varies from one type to another. This arrangement is found also in the light chains, which have about 220 amino acids: the constant zone characteristic of a type of immunoglobulin and the

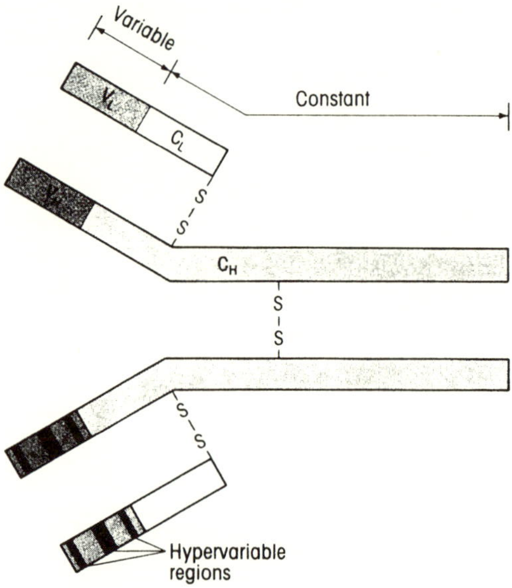

Figure 5. Typical structure of an immunoglobulin. The regions in black are the "hypervariable" regions; those in gray, the "variables." The heavy chains (C_H and V_H) are flanked by light chains (C_L and V_L), to which they are linked by disulphur bridges. (*Source:* I. Roitt, *Essential Immunology* [Oxford: Blackwell, 1989], 34.)

variable zone interspersed with hypervariable regions. Heavy and light chains contain long strips of identical sequences. It is the variable parts and especially the hypervariable parts that are responsible for the specificity of each type of immunoglobulin that is synthesized. The antibodies with the same variable regions but different types of constant regions in the heavy chains recognize the same antigens but have different roles in the immune response.

Two theories on the functioning of the immune system have been advanced: the germinal and the somatic theories. The latter has been developed mainly since 1965.

In the germinal theory, it was conjectured that the stem cells—the bone marrow cells that give birth to lymphocytes—must contain all the genetic information necessary to code all types of antibodies. A different gene would correspond to each protein, as is the rule in other domains of molecular biology, and each protein secreted by the lymphocytes would constitute an antibody for a single type of aggression. This formula would require that all the antibodies be coded separately, implying an extremely long genetic code.

The somatic theory, by contrast, supposes that the antibody is formed through combinations of fragments of genes, and this is the theory that has prevailed. In fact, the stem cells contain only a few fragmented genes. It is these genes, reassembled, that have the power to produce a multitude of different antibodies because of all their possible combinations. Simple, but it took nature to think of it. Figure 6 shows the DNA fragment in the germinal cells that corresponds to the light chain's protein code. It contains the zones coding for the constant segments and, scattered throughout the DNA, the sequences corresponding to the codes of the variable segments. The germinal cells give birth to the B lymphocyte stem cells. The B lymphocytes incorporate a constant sequence and an

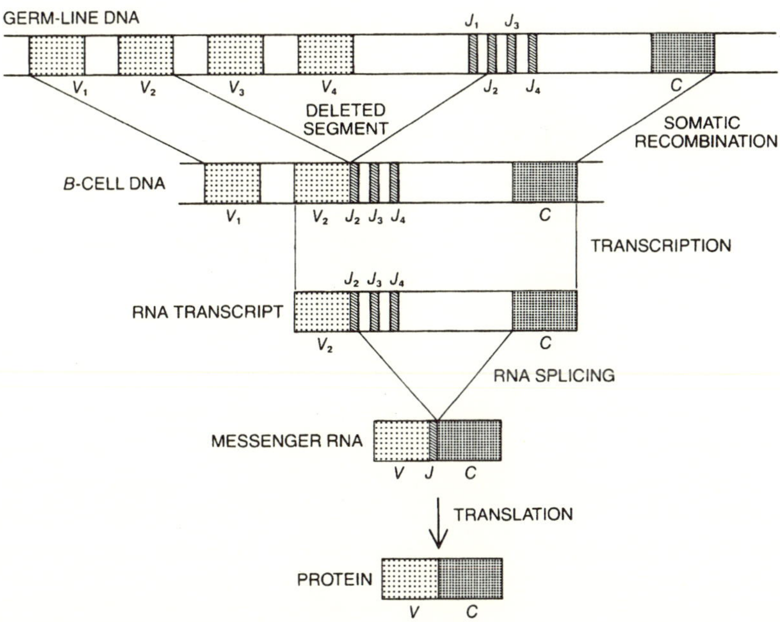

Figure 6. Sketch of the assembly of an antibody from diverse fragments of the genetic endowment of a germinal cell. This assembly takes place in two stages, shown here for a light kappa chain. In the first stage, enzymes assemble randomly chosen type *V* and *J* fragments, omitting the intermediary DNA. Here the segments marked V_3, V_4, and J_1 are omitted, so that the sequences V_2 and J_2 are joined in the DNA of a B lymphocyte. In the second stage at the level of transcription, the enzymes eliminate the intermediary portions of the lymphocyte B DNA, so as to produce an antibody molecule characterized by the assembly of a variable region *V* and a constant region *C*. (*Source:* S. Tonegawa, "The molecules of the Immune System," *Scientific American,* Oct. 1985, p. 127.)

associated variable sequence chosen from among all the germinal cell's possible pairs through a process that presents all the characteristics of randomness. In this way a very large number of different types of lymphocytes can be obtained, each of which can specialize in the production of a specific antibody.

Figure 6 also schematically represents the code of a light chain with its constant and variable parts. These sequences come from discontinuous zones on the DNA chain, separated by "introns," which are the parts of the DNA that are neither transcribed nor translated into amino acids in the construction of the final protein.[1] These partial sequences are reassembled to result finally in a constant segment (which is always the same) and a combination from among the approximately one hundred and fifty pieces of the variable part, in such a way that the mRNA that is constructed will code for a specific antibody protein. In the heavy chains there are several types of constant zones possible; there is also recombination for the construction of the variable part.

The hypervariable parts correspond to the junction points between the sequences. The junctions differ from one antibody to another, conferring even more diverse immunological properties.

The random recombination of the different sequences allows for an impressive number of types of antibodies, but their variety is even greater because of "somatic mutations," which affect the genome of the immune system cells during their maturation.

To summarize, we have in sequence the reproduction of the stem cells (in bone marrow), the production of differentiated B lymphocytes able to produce a type of chain, and the recombination of the fragments. All this happens before the lymphocyte encounters an antigen. Following that encounter, a reaction occurs if the antibodies correspond well with the antigens. The lymphocyte is stimulated by the antigen. All the lymphocytes that react to that antigen reproduce themselves, while the others remain on the sidelines of the immune reaction. Some among the B lymphocytes will become memory lymphocytes; others will proliferate to produce the needed antibody in great quantity.

During this time of replication, somatic mutations occur, a second important source of diversity. These mutations can result by chance in an even more perfect correspondence with the antigen and thus stimulate production of more specific antibodies. Somatic mutations affect only the variable and not the constant segments of the chains. It is estimated that the rate of mutation affecting these variable regions is around one hundred times higher than the overall average rate of mutation in the DNA of eukaryote cells (cells with a nucleus). This mechanism irresistibly calls for comparison with the process of artificial selection Darwin studied at length before constructing his theory of the origin of species.

From this qualitative description of the variety of immunoglobulins, let us pass to quantitative considerations. The DNA chain of the stem cells contains one hundred and fifty different sequences coded V (for "variable") and five sequences coded J (for "junction"), in addition to the constant sequences (C). It is the assembly of a J, a C, and a V that results in a gene coding a light chain. Therefore there can be $150 \times 5 = 750$ different light chains. The attachment of the V and the J can take place through ten different attachment zones. Thus, there are 7,500 different possible light chains, assembled in a random fashion. The possible variety is even larger for the heavy chains. These chains include a type of supplemental variable structure, called D (for "diversity"). There are eighty V zones, six J zones, fifty different types of D zones, ten types of attachment strings between the V and D zones, and ten types of attachment strings between a D and a J. Thus there are 2,400,000 possible heavy chains. Given that an antibody is formed of a light chain and a heavy chain, the combination of those two can produce a possible 18 billion different antibodies.[2] This enormous diversity is created essentially in two stages: either by working on the DNA during the cell's maturation or by excising and splicing the RNA before its translation into protein. The first stage, during which one can observe migrations of gene fragments on the DNA, is a phenomenon found only among the immune system genes.

How do T cells recognize their targets? The receptors present on the surface of these cells greatly resemble the immunoglobulins, discussed earlier for B lymphocytes. The type of code is the same for both types of lymphocytes. T lymphocytes have constant and variable regions, too, and they exhibit just as large a variety as do B lymphocytes. The special character of this type of lymphocyte consists in that it cannot combine directly with the antigen. Just as B lymphocytes are adapted to the battle against the antigens, which circulate freely, so T lymphocytes search for internally infected cells. The T lymphocyte is well adapted to its role, which is to kill these cells. It attacks and digests the cells that are already infected by the virus or the alien bodies. The infected cell attaches itself to the T lymphocyte in such a way as to trigger a so-called double key mechanism of recognition. The first key proves to the T lymphocyte that the cell it has encountered is indeed a cell belonging to its own body. This recognition of self by self takes place through identification, on the cell's surface, of the proteins characteristic of the major histocompatibility complex, which acts as a sort of identity card that all the cells of the human body possess. On the other hand, the lymphocyte also recognizes the antigen's signature on the infected cell's membrane. This second key confirms that the cell has been contaminated by an antigen for which the lymphocyte produces precisely the specific antibody. It is only in the presence of these two signals that the lymphocyte can do its death-dealing work.

These mechanisms have been understood for only a short time and are still being actively researched. There are two competing theories as to the precise mechanism of the double key. One theory proposes two different sites, the first charged with recognizing the major histocompatibility complex and the other charged with recognizing the antigen. But the two stages might occur in succession rather than simultaneously in separate locations. The antigen would link to the cell to form a complex, and it would be that complex that would be tested at a particular T lymphocyte location.

IMMUNOLOGY, CHANCE, AND THE ACQUIRED IMMUNODEFICIENCY SYNDROME (AIDS)

AIDS, as everyone knows today, results from the infection of the human body by a virus, or more precisely a retrovirus. A retrovirus is a virus whose nucleic acid, of the RNA type, can transform into DNA inside the cells it infects. This DNA is then incorporated into the cell's genetic endowment. Thus, when the DNA of the cell expresses itself, it produces copies of the viral RNA. In this way the infected cell becomes a true virus-multiplying machine.

One unique aspect of the AIDS virus is that it preferentially infects the amplifier T lymphocytes, still called T4 cells. This type of white cell is specifically charged with stimulating the reproduction of the B lymphocytes and the specific antibodies that they produce. Clearly, then, through an extraordinary piece of bad luck, this disease is especially difficult to fight. From the very beginning, the virus jams the organism's normal autodefense mechanism.

How does the virus preferentially attack the T4 lymphocytes? The surface of the virus carries a receptor that recognizes a characteristic protein of these lymphocytes, the CD4 protein. Once it recognizes this protein, the virus anchors itself to the lymphocytes and then merges with them, allowing the nucleic material it contains to do its work (by using the enzymatic apparatus of the cell it infects).

The infected cells do not actually die, or at least they do not die right away. AIDS is characterized by a surge of open multiplication of the virus in the days following the infection. This stage is followed by a long period of dormancy during which a slow transformation of the infected cells seems to occur. At the beginning the infection is fought by the production of antibodies that manage to reduce the numbers of viral cells present considerably. It is this surge of antibody production that is detected in the seropositivity tests. The numbers of viral particles in the blood return to low levels without being annihilated. Little by little, however, the remaining infected lymphocytes transform themselves and bring death not only to themselves but to many neighboring lymphocytes (by way of a secondary process of contamination that I will not describe here). The immune defenses are whittled away, and the number of viral particles grows as the number of living T4 lymphocytes decreases. Soon the depressed immune system becomes incapable of marshaling itself against the opportunistic illnesses that generally mark the appearance of full-blown AIDS.

■

This description of the mechanisms presiding over the functioning of the immune system in defense against all attacks on their identity and integrity may seem cumbersome. For anyone wanting to reflect on the role of chance in the living world, however, there is no way to avoid the effort. Once accomplished, this effort leads to the marvelous discovery, at the heart of the most intimate defense mechanisms of life, of strategies that are greatly reminiscent of the fundamental ideas of Darwinism. Darwinism, remember, has two essential ingredients: (1) variations that present all the characteristics of the industriousness of chance, at least as regards the second ingredient, which is (2) selection ruled as much by elements external to the organism as by internal conditions, if not more so. The production of antibodies conforms perfectly to this outline. Everything is to be found there: variations by chance from a reservoir of elements, selection of the combinations most likely to fulfill the functions of defense given the aggressors present in the environment, and finally, the mechanisms of memory necessary to make the system evolve. A modern exposition on Darwinism could begin by reporting on immunological mechanisms.

And there are more surprises in store for us. An examination of the mechanisms of the construction of the central nervous system and of the functioning of our brains will reveal new Darwinian processes bearing the clear fingerprints of natural chance.

NEURAL DARWINISM; OR HOW ARE NEURAL NETWORKS FORMED?

9

The complexity of the central nervous system is another illustration of chance's opportunistic role in life. We know that the human genome contains only a few billion "letters" (nucleotides), and it is estimated that fewer than 100,000 genes are truly expressed (translated into proteins) during the formation of an adult organism, a number clearly insufficient to cover all the necessary details of that organism's organization. This is especially obvious with the nervous system. The nervous system alone contains about 100 billion neurons. These cells are widely interconnected. Many neurons receive information from several thousand other nerve cells. The number of synaptic contacts between neurons in the central nervous system is estimated to be more than 10^{14} (100 trillion). It is thus literally impossible that the linkage of such a system could be mapped out in detail, inscribed in the individual's genetic endowment, as in the wiring schematics of a television set.

AN ARCHITECT WHO DOESN'T SWEAT THE DETAILS

Nature thus does not use a blueprint. We still do not understand very well how it manages to reproduce the same communication centers and similar wiring from one individual to another. The embryology of the central nervous system is still being actively researched. The American biologist Gerald Edelman (Nobel Prize, 1972) emphasized the role of surface proteins that recognize neighboring cells and determine, from among a group of undifferentiated embryonic neurons, the list of different nerve centers in which the neurons acquire an overall homogeneity while markedly differentiating from the neurons of the other nerve centers. The cell body, or soma, takes different forms: basket, star, granule, and so on. It establishes dendritic extensions (the roots with which cellular bodies "absorb" incoming information) and an axonal trunk charged with the outbound transmission of the nervous impulses created by the neuron (fig. 7). Once hav-

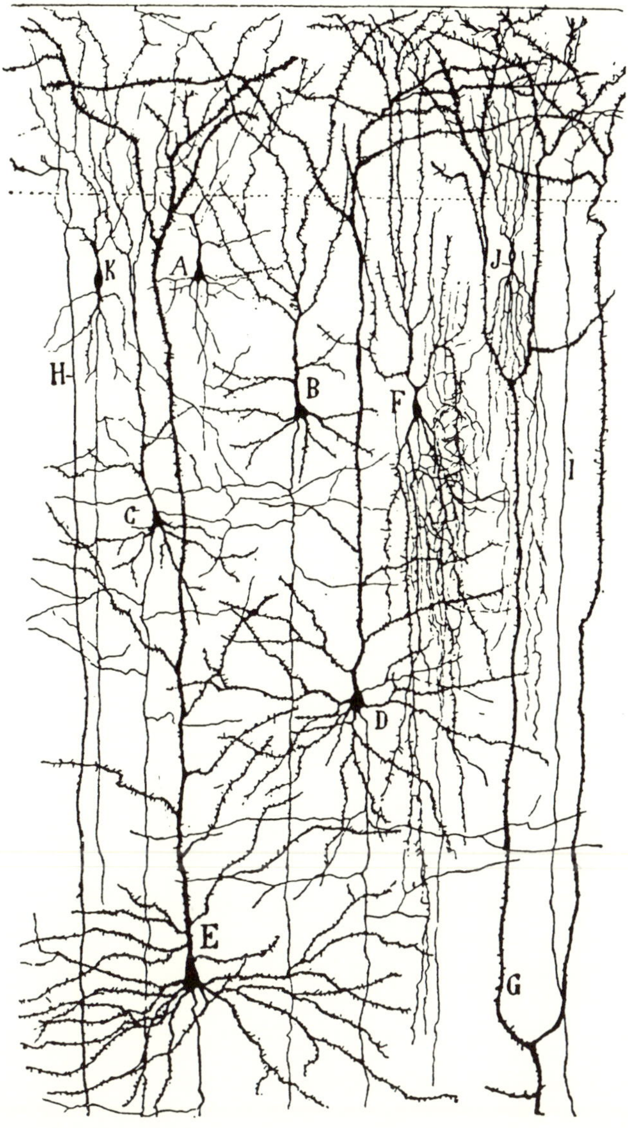

Figure 7. Some examples of neurons as they appeared under the microscope of the Spanish anatomist Santiago Ramón y Cajal around 1910. For example, the B neuron is composed, from top to bottom, of a diversified dendritic arborization and has many "dendritic spines." The soma of this neuron shelters the nucleus. The axon extends from the lower part of this body, a fine filament that goes well beyond the boundaries of the figure. For a neuron at rest, the difference in the ionic composition between the neuron's internal and external environments establishes a difference in electrical potential between the two sides of the surrounding membrane. The internal environment reaches about –70 mV relative to the external environment. (*Source:* S. Ramón y Cajal, *Histologie du système nerveux de l'homme et des vertébrés,* rev. ed. [Madrid: Consejo Superior de Investigationes Cientificas, 1955], 2:544.)

ing arrived in the neighborhood of its target cells, the axon ramifies, often branching out like a bush. Some of the terminal fibers are no thicker than 0.2 microns in diameter.[1] Information is transmitted from one neuron to another across synaptic junctions. The terminal fibers of the axons often carry strings of synaptic "boutons" (buttons), called "en passant" (passing by), extending to the final bouton. Finally, the neurons of a particular functional group are differentiated from the neurons of other groups through the type of chemical neurotransmitter that the synaptic boutons release when traversed by a nervous impulse.

How do these threadlike extensions direct themselves toward their target cells as they are growing? Biologists talk about the "growth factor" and "growth cones" in regard to this subject. Chemical substances would thus pilot the axonal branches somewhat as, in France, it was hoped that special planes could be developed that would be able to locate oil fields. We might as well admit that the mechanism is still largely unknown. Remarkable experiments have taken place at the Institute of Neurosciences in Paris to determine how, during the development of young kittens, the nervous fibers of the visual system can make spectacular turns of almost ninety degrees (characteristic of the optic nerve) when they leave the retina to colonize the lateral geniculate body in the thalamus, doing so when and only when the fibers come from the side of the retina that receives visual stimulation from the contralateral visual hemisphere (so that the right lateral geniculate body receives fibers carrying visual information relative to the left visual hemisphere and vice versa). Figure 8 illustrates this astonishingly blind, groping construction of the visual system. In the young animal the synaptic contacts established during embryogenesis are superabundant, and most of them will eventually disappear. It is thought that the use or nonuse of nervous pathways determines the fate of the nervous connections at this stage of development.

At the end of this critical period or of successive ones, the general map of the nervous system, with its hierarchy of centers, is definitively established. It allows a great deal of individual variability. We know, for example, that despite the great physical resemblance of identical twins, their central nervous systems are not configured identically. This was resoundingly confirmed by a microscopic examination of the ramifications of neurons observed in the *Daphnia magna,* a type of small freshwater crustacean that reproduces through parthenogenesis; consequently, all descendants of a single individual have an absolutely identical genetic endowment (fig. 9).

BILLIONS AND BILLIONS OF NEURONAL CONTACTS

Microscopic examination of a small region of nervous tissue, without using selective dyes, can plunge the observer into confusion that can approach despair.

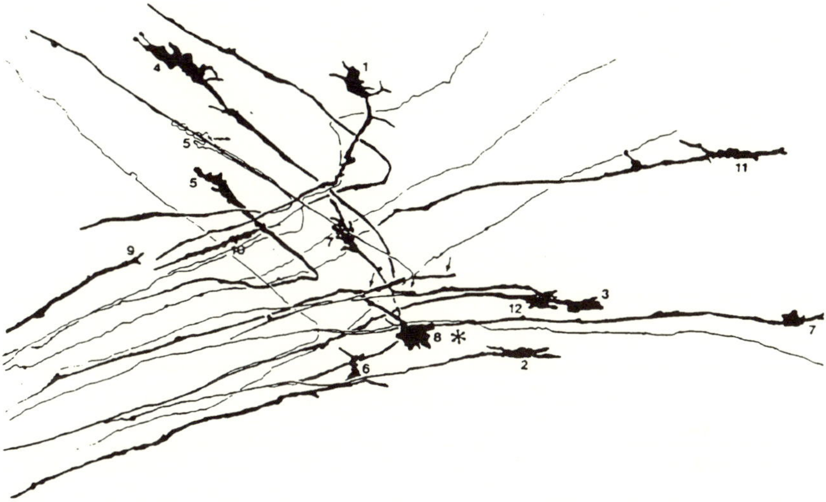

Figure 8. Construction of visual pathways in the young mouse. Their groping progress from the retinal ganglia does not impede about half the optical fibers from correctly making a sharp turn toward the visual nuclei of the thalamus. (*Source:* P. Godement, J. Salaün, and C. A. Mason, "Retinal Axon Pathfinding in the Optic Chiasm: Divergence of Crossed and Uncrossed Fibers," *Neuron* 5 [1990]: 176.)

How can one orient oneself in that tangle of fibers, cell bodies, sheaths, fiber-to-fiber contacts, fiber-to-soma contacts, contacts to contacts, tangled together in no apparent order? It seems that we must stop thinking of the information circulating through the nervous pathways as messages circulating through a network of telegraph lines. The nerve centers are composed of *assemblies* or *groups* of relatively homogeneous neurons, equivalent and interchangeable, that can receive information to be transmitted in parallel. We are faced with not lines for transmission of information but pipelines, or rather tangled skeins of unbelievable complexity,[2] enhanced further by the diverse nature of the contacts that a given cell receives. Some synaptic contacts are excitatory; others, inhibitory. Some neurons use as messengers the chemical intermediary of glutamic acid, whereas others use norepinephrine, dopamine, acetylcholine, serotonin, g-aminobutyric acid, or other chemical molecules (we know of more than twenty). Synaptic contacts use neurotransmitters to transmit excitation or inhibition signals. Other chemical compounds called "neuromodulators" control the downstream neuron's responses. Each compound has its own characteristics. It acts on different receptors of the downstream cell, possesses its own response time and elimination time, and so on.

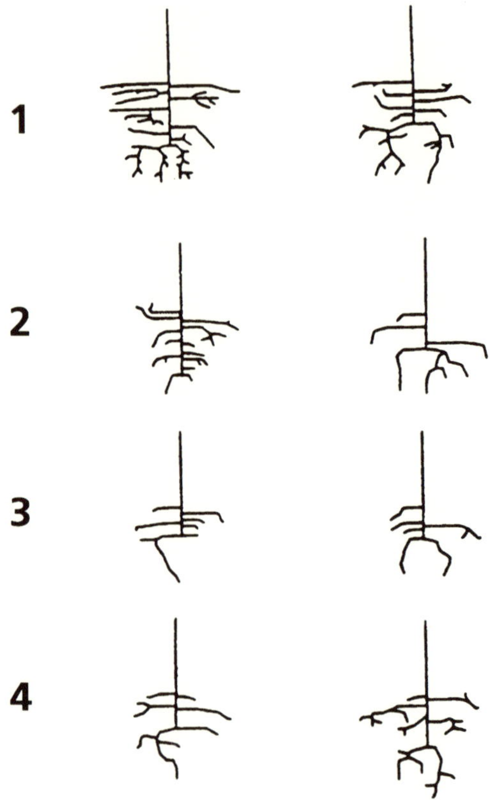

Figure 9. The homologous neurons of four specimens of *Daphnia magna* show differences in many of the details of their topographic organization, even though the genetic endowments of all these individuals are absolutely identical (for each individual the two neurons corresponding to the right and left sides of its nervous system are represented). (*Source:* E. Macagno, V. LoPresti, and C. Levinthal, "Structural Development of Neuronal Connections in Isogenic Organisms: Variations and Similarities in the Optic System of *Daphnia Magna*," *Proceedings of the National Academy of Sciences, USA* 70 [1973]: 57–61.)

At the heart of each population of neurons forming a nerve center, electrical activity can determine the appearance of specialized subassemblies that Edelman calls "secondary repertoires." These neuronal groups are true nodes along the communication routes of nervous information. The network they establish with one another depends greatly on external stimuli. With each stimulus, different groups of neurons are excited. The groups that respond together to a given stimulus and that are linked by synaptic connections tend to reinforce their solidarity—the efficiency of their synaptic connections increases. In contrast, the connections between neurons that are never or rarely excited together become less and less sensitive and can even eventually degenerate and disappear.

As early as 1949, the great Canadian physiologist and psychologist Donald Hebb already suspected that the efficiency of synaptic connections is malleable thanks to some unspecified physicochemical mechanisms and that this property is the basis for the plasticity of the overall nervous system—that it is the key, in particular, to the phenomena of learning and memory. Not until fairly recently were researchers able to develop the means of studying the detailed physicochemical mechanisms brought into play by this plasticity. Current studies are revealing a great diversity in the nature of the general properties of synaptic plasticity: long-term potentiation, long-term depression, and short-term facilitation,[3] with effects that may last from some milliseconds to days. The involvement of these mechanisms in memory formation is, however, still a matter of debate.

THE DEVELOPMENT OF NEURONAL NETWORKS

Once the synaptic contacts have been established along the visual routes, the visual system begins to function in the embryo. Even in the darkness of the uterus and in the absence of external stimulation, it is traversed at more or less regular intervals by electrical discharges that certainly play a role in the construction or the consolidation of specific neuronal circuits. It is the same with other sensory systems. It is not necessary, in fact, that the "bursts" transported in this way obey an exact, predetermined code that is dictated by the genetic endowment. These codes are more probably selected from a largely random mode of discharge when the nerve cells begin to fire. Nature thus would profit, once again, from chance variations to "invent" a code and a meaning. Whatever the case, the spontaneous activity of the neural circuits is indispensable to the correct development of the cerebral architecture. For example, if the spontaneous activity of the embryo's visual pathways is inhibited by chemical substances, aberrant pathways can result, and the adult may even be permanently blind.[4]

Similarly, it is possible that the transmission of information itself, in the heart of the already consolidated adult central nervous system, depends on the chance emission of such "bursts of spikes." If, by chance, the particular temporal structure of some of these bursts were "recognized" by target neurons (because of the particular configuration of the synaptic contacts between the two neurons), the latter could form, for a time, a privileged channel of information transmission, and the bursts that the first would transmit to the second would thus be promoted to the rank of "messages." Such a process might thus account for the construction of effective circuits hidden among the diversity of anatomically possible networks, individualized by the temporal structure or the synchronicity of the nervous discharges that traverse them.

FLUCTUATIONS IN THE NEURONAL SIGNAL . . .

Let us examine what happens when a "message" is transmitted from the axonic terminal bouton of one neuron to a dendritic sensitive zone of the downstream neuron. The electrical signal that runs through the axon of the first neuron stimulates the expulsion of globules of neurotransmitters when it arrives at the presynaptic bouton,[5] which in turn immediately expand into the intersynaptic cleft. These neurotransmitter molecules attach themselves to specific molecules on the dendritic wall of the second neuron: the receptors. The attachment of the neurotransmitter molecules to these particular sites causes ionic channels to open, somewhat as the insertion of magnetic cards opens garage doors. The stream of ions that then surges into the postsynaptic zone of the downstream cell[6] causes the appearance of an electrical signal, the postsynaptic potential. This potential, when it reaches the cell body, is added to other postsynaptic potentials (positive or negative) formed in other synapses scattered along the dendrites or the body of the neuron. When all the combined potentials reach a certain threshold, an avalanche of ions is unleashed at a special point on the neuron between the cell body and the head of the axon, creating an electrical "shock" that propagates like a wave along the axon. This is the so-called nervous impulse, or the *action potential.*[7]

All the action potentials traversing a given nerve fiber have the same amplitude and the same temporal variation.[8] In the context of this study of chance, note to begin with that a synaptic bouton signaled by the action potential does not always open a neurotransmitter vesicle. The release generally occurs by chance, with a certain probability. In general, however, the linkage between two neurons is composed not of one but of several tens of synaptic contacts (fig. 10). The number of synaptic vesicles opened because of an action potential produced by the first neuron (and thus the "postsynaptic" potential induced in the second neuron) varies according to chance, obeying the binomial law characteristic of independent events.[9] It has been possible to demonstrate clearly the quantified fluctuations associated with these "chance drawings," just as it has been possible to display fluctuations over time in the number of open or closed postsynaptic ionic channels.

The nervous system accommodates itself to these fluctuations in two ways. On the one hand, two contiguous neurons form just one synaptic contact only in exceptional cases. Rather, the contacts form tightly grouped clusters of a few units or of some tens of synapses. A single neuron can project several different clusters onto the dendritic network or the cell body of a single target neuron, as is the case in contact between the neuron coming from the retina and the neuron of the lateral geniculate body (a relay center for vision, situated at the center of the brain in the thalamus between the eye and the visual cerebral cortex) shown in figure 10.

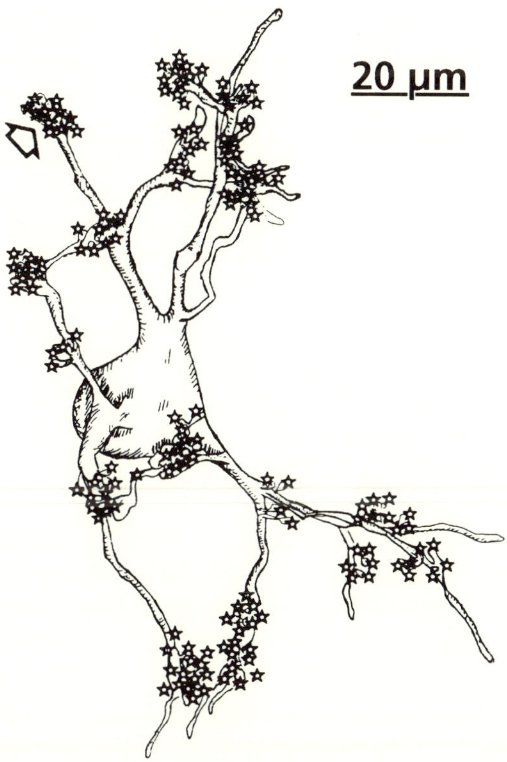

Figure 10. Placement of the synaptic contacts between the terminal fibers of a neuron from a retinal ganglion and its target cell in the lateral geniculate body. Note the disposition of these contacts in separate clusters, evoking the possibility of a sort of very precise spatiotemporal signature of the passage of information from one neuron to another. (*Source:* J. E. Hamos, S. C. van Horn, D. Raczkowski, and S. M. Sherman, "Synaptic Circuits Involving an Individual Retinogeniculate Axon in the Cat," *Journal Comp. Neurology* 259 [1987]: 176.)

The arrangement of the synaptic contacts in clusters has an obvious advantage. The law of large numbers can then ensure that despite fluctuations of the signals delivered to each synapse, the total synaptic potential delivered remains fixed within fairly narrow limits. The dispersion into distinct clusters can play a role in the temporal coding of nervous information, as I have been proposing for several years on the basis of B. L. Strehler's original suggestion.[10]

The second way in which the nervous system accommodates itself to the fluctuations that characterize its functioning lies in the high level of parallelism of its avenues of communication. As already noted, information seems not to traverse precisely defined, specialized routes reduced to a single nervous fiber but rather to concern groups of neurons and bundles of fibers.

It has proved possible to estimate the importance of these groups of equivalent neuron populations from observation of the detailed structures in the cerebral cortex called "microcolumns." Edelman estimates that each individual group of neurons forming the elementary unit of reference to which the coding mechanisms apply must contain fifty to one hundred neurons. There, too, very rapid transitions of the avenues of communication from one fiber to another must contribute to rendering the process of information transmission true and sure.

. . . DO NOT KEEP IT FROM BEING AN EXTREMELY PRECISE TOOL

Whether in this way or in another, the inescapable conclusion is that the nervous system constructs a logical system of incomparable precision from imprecise and largely stochastic elements.

Precise the nervous system certainly is, to an absolutely unbelievable degree, taking into account the time spread of elementary events or action potentials. I have mentioned that these individual signals possess a characteristic temporal lifespan of several milliseconds, with a narrow peak of depolarization more than one millisecond long (fig. 11). And yet, many examples, such as the study of bat sonar or the owl auditory system, have demonstrated that the brains of these animals can detect and measure time delays between sensory signals certainly less than a tenth of a millisecond in length.[11] Thus, beyond using the random fluctuations present at different levels in individual nervous phenomena, the brain seems to make use of the law of large numbers. It must also take advantage of the enormous variety of the "messages" composed of a burst of nervous impulses (due to the possible combinations of the time intervals separating the impulses within the burst) to transport precise messages efficiently. But this last point has not yet been formally proven.

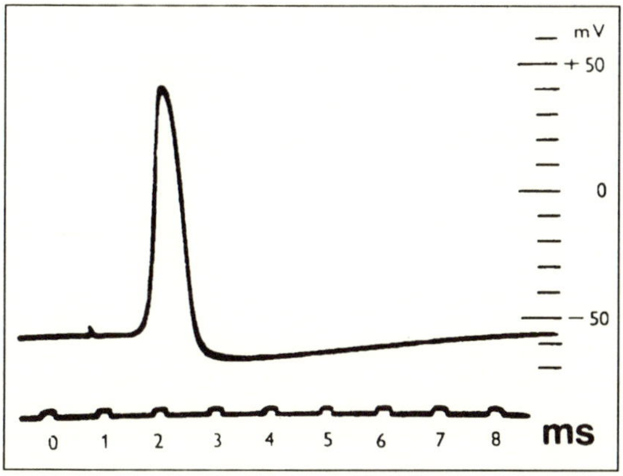

Figure 11. Recording of a typical action potential on a squid axon. Studies conducted by the Britons Alan Hodgkin and Andrew Huxley since 1939 allowed them to propose, for the first time, a precise mechanism for the active (nonattenuated) propagation of this impulse along the length of the axon. They were rewarded with the 1963 Nobel Prize. (*Source:* P. Baker, A. Hodgkin, and T. Shaw, "Replacement of the Protoplasm of a Giant Nerve Fibre with Artificial Solutions," *Nature* 190 [1961]: 885–87.)

A BENEFICIAL NOISE

Performances of this type seem even more extraordinary given that the functioning of the nervous system takes place against a backdrop of continuous noise. Not only do individual synaptic contacts not function each time they are solicited, but the neurons of the central nervous system are subject to spontaneous discharges, independent of all external stimulation. There is constant interference, as if the neurons were "talking" to one another over telephone lines that were continuously bombarded by a storm of static.

One of the most remarkable contributions of the theory of neural networks these past few years is the demonstration that noise *is necessary to the correct functioning of such intelligent systems.* The reason for this requirement is that these networks possess not just a single stable state but hundreds or thousands of possible stable states, each characterized by a relative minimum of their "energy." A network of neurons without noise could not spontaneously go from one minimum to another. With each stimulation it would be booby-trapped by the first minimum it encountered during its evolution. Only the noise allows it to explore all the transitory local minima before finding the absolute minimum in which it must install itself, depending on external constraints and stimuli. It is in fact this state that constitutes by definition the neural network's "response" to specific external stimuli.

Thanks to this plasticity of neural networks, it is possible to construct "neuromimetic" computers whose functioning is inspired by the principles at work in biological brains made up of neurons but whose basic elements are semiconductor chips. Such machines possess the properties of "associative" memory (addressable according to content) and "robustness" (slightly different stimuli can bring about the same response), which also characterize biological nervous systems.

Anecdotally, 1990, the hundredth anniversary of the coining of the word *neuron* (by Waldeyer),[12] also marks the hundredth anniversary of Edouard Branly's invention of the "coherer," the first instrument capable of detecting radio waves (using iron filings to do so). Branly, it is said, was fascinated by the disordered structure of the nervous system and wanted to imitate it with his detection device, whose efficiency remains a mystery to this day.

FROM NEURAL DARWINISM TO THE DARWINISM OF IDEAS

Gerald Edelman and Jean-Pierre Changeux have formulated similar proposals about the mechanisms of selection that might preside over the brain's construction and functioning. At the time of the embryo's formation, the brain applies general and simple rules of construction, inscribed in the genome. These rules specify the nerve centers to be constructed and determine the specific type of neurons of which each center must be composed. The homogeneity of the neurons of similar function in the same center (same general structure, same type of nervous targets, and same neurotransmitters used) and the differentiation of the centers from each other are ensured, it seems, by contact messages, from nerve cell to nerve cell, in particular by surface proteins such as the CAMs (cell adhesion molecules) and the SAMs (substrate adhesion molecules). This phase of construction takes place only during a critical, genetically determined period. The nervous connections in the brain then undergo a rapid branching out that varies from one individual to the next. Superfluous connections that prove unusable degenerate and leave behind a well-organized brain with a clear architecture. Thus one has been able to estimate that, in the cat, the number of axonic fibers of the optic nerve drops from about 430,000 on the fiftieth day of gestation to only about 160,000 at birth.[13] The degenerative process of useless connections in the visual system thus begins even before that system serves its purpose: to see. This process is probably controlled by the spontaneous impulses originating in the embryo's retina. But birth does not necessarily mark the end of this critical period. In the prefrontal cortex of the macaque monkey, the number of synaptic contacts (by cubic millimeter of cortex) drops, on average, from about 800 million during early infancy to 300 or 400 million after puberty.[14] In the brain, work units are formed from groups of neurons connected to one another, homogeneous but individualized. They form circuits or networks that constantly change configuration, in some tens of milliseconds, according to the type of nervous excitation they receive, and more precisely, probably as a function of very precise temporal signals immersed in the flood of signals apparently randomly emitted by the excited neurons. According to Edelman and Changeux, these changes are driven by a variation-and-selection mechanism similar to the classical Darwinian plan, from which comes this theory's name, "neural Darwinism."

In this vision of the brain's workings, where is mental representation? For Edelman, mental representation emanates from the correspondence between several networks, each of them specialized in a different primary task. For example, one would be charged with detection of the external physical form of the perceived object, and another would be charged with analyzing its geometric topology (e.g., the number of handles on a cup). Or again, the first circuit would specifically detect the orientation of the segments composing the contour of a figure, and another circuit would comprehend the direction of that figure's motion. According to this author, the topographic maps that develop in the central nervous system and that are specialized for each of these tasks must maintain exchanges and reciprocal connections between one another. It follows that the representation of an "object" in the brain is at least *dual* and probably multiple. Edelman insists especially on the importance of "reentries," or reciprocal connections between populations of neurons, which abound in the brain and whose function is still poorly understood.

Changeux bravely proposes that the Darwinian tapestry that applies to the two preceding stages (selection of the assemblies of neurons and reciprocal switching of the networks they define) applies also to the formation of ideas. Soon after the intense and burgeoning activity of the unconscious, translated into terms of physiology by the simultaneous activation of multiple networks, would come the selection and crystallization of a pertinent idea, or at least one judged to be pertinent, which would then alone be elevated to the level of conscious perception. According to some researchers (Benjamin Libet), both incomplete and conscious representations could be produced by the same centers and the same networks, but the former would be extremely unstable, whereas the latter would require a greater stability over time.[15] This would explain why a single center of conscious representation in the brain has never been located. The external world would be the most permanent, lasting re-creation, from among all the populations of possible worlds, of the most stable vision, the winner of a typically Darwinian competition for survival.

Within this general framework, the Darwinian outline of variation and selection would apply finally to four successive levels. First, it would apply to the construction of the neural architecture, thanks to the potential variability of the embryonic nerve cells and to the superabundance of their connections. Second, it would preside over the selection of the neuronal assemblies and the most efficient networks for dealing with sensory stimuli, the conscious representation and preparation of motor activity, thanks to the diversity of the nodes and of the possible alternative circuits. Third (according to Changeux and Dehaene),[16] the same type of plan would apply also to the sorting, from among the multitude of ideas and their organization in subconscious logical correspondences, of single ideas that are sufficiently stable or efficient to merit being elevated to the rank of ideas represented in the consciousness. And finally, of course, the original mechanism of variation and natural selection, working at the level of the entire individual, is at the service of the individual and the species survival program.

This last selection, the highest in the hierarchy, would, because of its position, control all the others. Despite fundamental differences in the types of variations and the mechanisms of selection, not a single one would be able to oppose the interest of the individual's or the species' survival.

But it is here that the Darwinism of ideas seems to me to cause a problem, since it is necessary that the mechanisms that ensure the triumph of the most stable perceptions ensure at the same time the most pertinent vision, or the vision that is most apt to ensure the survival of the individual. Such a conception is hard to reconcile with the imperatives for speed in the expression of ideas and of perceptions. The survival of the individual often demands the most rapid response, which could not allow time for laborious selections between unconscious perceptions. In addition, when it comes to stimuli, everyone knows that the most fleeting of impressions are not necessarily the least significant or the least important.

Edelman thinks that the more certain neural Darwinism is (at the level of both "permanent" primary architecture and the short-lived secondary assemblies, or "secondary repertoires"), the less probable is the Darwinism of ideas (or mental Darwinism). Language and conceptualization, he thinks, create permanent pigeonholes in the brain that, once created, are there to stay. Discourse and the creation of ideas are thus characterized by a combination of permanent elements in the brain, much more so than by trials, by variations, and by Darwinian-type selections to which unstable and ephemeral neural networks would correspond.

■

As ever-present as it is in the highest manifestations of life, natural chance does not limit its realm of influence to life. It is time now to reexamine chance's role in inanimate phenomena, from the atom to the cosmos. In fact, physical entropy shares many of the characteristics of natural chance already seen in the sphere of the living. And entropy is not just the simple measurement of physicomathematical disorder to which some have tried to reduce it.

PART

3

PHYSICAL CHANCE

ENTROPY, ORDER, AND CHANCE

10

HEAT NEVER TELLS WHENCE IT CAME

Physicists excel when it comes to prediction. They are given the *initial* state of the system and asked to predict what its state (e.g., the position of a planet in the celestial sphere) will be at a *later* moment. The equations of classical physics, such as Newton's equations of motion, allow the configuration of a physical system to be calculated at any instant as long as its state at a given instant and all the forces acting on it are known. But the equations used to make predictions can also serve in reverse, for retrodiction. Thanks to these equations, for example, the planetarium of the Palais de la Découverte can re-create the configuration of the skies at the time of the pharaohs or in the age of the dinosaurs.[1] The present state of the skies is taken as the initial state, but the velocities of all the celestial objects are reversed. This allows us to see the sky rerun all its previous configurations.

This indifference of the equations to temporal direction makes them inefficient in accounting for the simplest experience of the beings that people this world, that of temporal irreversibility. For classical physics, nature can unmake all that it makes. The filmed sequence of a high diver, run in reverse through the projector (beginning at the end), shows the disturbance of the water expelling the diver from the wave and taking him back, through astonishing twists and turns, onto his toes at the edge of the board. Such a sequence is not shocking when viewed through the laws of mechanics. It represents only what would be observed if, after the dive, the same configuration of water molecules and of the diver's body were preserved, but the velocity vector of each were reversed.

There is in fact a physical law, known since the beginning of the nineteenth century, in which this symmetry between prediction and retrodiction does not hold. This is Fourier's law on heat diffusion. Given a bar heated at one end, Fourier's law describes the future distribution of that heat throughout the entire bar. But here the experience with the diver cannot be repeated. Under no circumstances can Fourier's law predict that the uniformly warm bar would change back into a bar that would become hot at one end and cold at the other. Which end would be hot? And why would it be this side rather than the other? Fourier's law was explicitly constructed for prediction and is notoriously inept in retrodiction. It is said that it is not symmetrical through "time reversal," which is a condensed but awkward expression (who can reverse time?).[2] But this peculiarity was not noticed, either by Fourier or by his immediate followers.

It was not until the middle of the twentieth century that temporal irreversibility was associated with a precisely defined concept in physical theory. This concept is that of *entropy,* used today not only in thermodynamics but also in chemistry and biology and by mathematicians, computer scientists, and even sociologists. It goes without saying that these different categories of specialists do not always give this word exactly the same meaning, other than through the use of analogies that are sometimes insufficiently grounded or explained. The nonspecialist might have difficulty deciphering its meaning.

ENTROPY STANDS IN OPPOSITION TO *ENERGY*

When the concept of entropy was created, physicists were still fascinated by the then recent discovery of the general law of the conservation of energy. Energy is conserved in our world, and the changes we observe are only manifestations of its transformation from one form to another: kinetic energy as well as mechanical, electrical, chemical, or caloric. That understanding brought the realization that muscular energy, too, is only a form of chemical energy acting in living cells. But the statement of the general law of the conservation of energy is also the affirmation, par excellence, of the absence of true change in the world. It gives the final touch to the nontemporal perspective of the laws of mechanics. The most astounding and precise demonstration of this fact was given by the German mathematician Emmy Noether in 1918. The theorem she bequeathed us demonstrates specifically that every system whose description in mechanics is indifferent to any displacement in time, even infinitesimal ones, is a system in which energy is conserved, and vice versa. In other words, the conservation of energy applies each time a system's trajectory (in real space and in its phase space)[3] is the same whether the experiment is conducted now, a second from now, or at any other moment. This theorem, established in classical mechanics, has since been extended to quantum mechanics. It supposes, of course, that the system in question is isolated, or at least that the external forces acting on it do not vary across time, either in intensity or in direction. This theorem expresses the idea that—except perhaps for describing cosmological processes for which the evolution of the universe must be taken into account—the world in which we live is one of harmony and indifference to time. The concept of energy and its law of conservation are the exact expressions of this harmony.

The world does change, however, and objects, like beings, age. This is why Clausius placed "entropy," symbol of that which truly changes even in cyclical situations where all things seem periodically to regain the same values, in opposition to "energy," symbol of everlastingness. Clausius prescribed how to calculate—rather than measure—the changes of a system's entropy. These changes are linked to heat exchanges between the system and its surroundings.[4] The classic example for which it is easy to calculate the growth of entropy is that of a steam engine, which works by absorbing heat from the firebox and returning it to the condenser. The machine's complete cycle involves admitting steam (produced in the boiler) into the cylinder, its expansion and production of work, its expulsion toward the condenser, and finally the return of the condensed steam into the boiler. At the end of the cycle, the steam is in exactly the same physical state as it was at the beginning, but during the cycle entropy was dispersed into the environment.

Clausius's original work was on "isolated" systems, since it is only in these systems that one can plainly demonstrate (on the basis of Carnot's principle, which states that heat always spontaneously travels from hot bodies to cold bodies and never in the other direction) that the entropy of a system can only grow (or at the least remain constant) over time. But isolated systems are artificial limits, unlikely if not impossible in the real world. In practice, physical systems always interact with the external world to some extent, even if only because there is no shield against gravitational influences. The true meaning of the Clausius-Carnot law is that, beyond the exchanges of energy and heat that a system might have with its external environment, internally it can only produce entropy and not destroy it. Thus the law of increasing entropy must be refined. If a system has entropy S, then every elementary transformation changes that entropy in such a way that it can always be decomposed into two terms:

$$dS = dS_i + dS_e,$$

where the first, the internal production of entropy, cannot be negative. It is at the heart of this term that, phenomenologically, the root of temporal irreversibility is to be found. This positive (or at the most null) term is the price that a material body or a physical system must pay for the right to exist in this world, which is universally marked by the arrow of time. If the second term of this expression, measuring exchanges with the outside, is null, as is the case with isolated systems, Clausius-Carnot's original law is uncovered anew.

In the world that surrounds us, there is thus an incessant production of entropy. One of the most common mechanisms of this production is the emission of photons, because photons usually carry a large amount of entropy. Indeed, in contrast to material particles (constrained by different laws of conservation), photons can appear or disappear almost freely, as long as the total energy is preserved. The dissipation of heat causes the proliferation of "soft" photons (of lesser energy and frequency), and this proliferation is accompanied by the creation of entropy. Thus, the Earth is a source of increasing entropy, since it absorbs energetic rays (in the sense that we speak of energetic food) from the sun and returns to space a much larger number of infrared photons, of a lesser energy (fig. 12). Considering the universe as a whole, by far the most of its proven entropy is supplied by thermal background radiation, that enormous ocean of photons at 3°K in which the entire universe is immersed and that is the visible remainder of the violent events[5] that formed our current universe about 15 billion years ago. The probable existence of black holes could modify this view on the production of entropy in the universe, since black holes are also great purveyors of entropy.[6]

BOLTZMANN LINKS ENTROPY TO CHANCE . . .

I must insist on one fact. Reluctantly, because it was completely contrary to the research program he had envisioned, in the 1870s Boltzmann established a relationship between entropy and chance. In fact, the young Boltzmann dreamed only of dismantling the workings of temporal irreversibility by using the tools of classical mechanics. According to him, achieving this goal required showing that the entropy of the thermodynamicists was merely a simple mathematical function of the position and velocity of each molecule composing the system under study. He had partially succeeded in this program when he unveiled the H theorem in 1872. This theorem shows that a certain function of the distribution densities (i.e., of the dispersion) of the positions and velocities of the molecules of a volume of "perfect" gas, called the H function, can vary in only a single direction under the influence of the collisions between molecules. These elastic collisions obey the laws of mechanics, which are time reversible, whereas the H function is unidirectional, like entropy, which can only grow in such an isolated system. Some years later, however, shaken by the serious objections presented to him, he recognized that his H function is not truly a function of the real positions and velocities of particles; rather, it is a function only of the laws of probability that govern them. As a consequence, the growth of entropy was not mechanically demonstrated—it was only shown to be extremely probable.

In addition, so as to be valid, the H theorem presupposes a particular initial condition called *molecular chaos*. Boltzmann allowed as an elementary, obvious fact

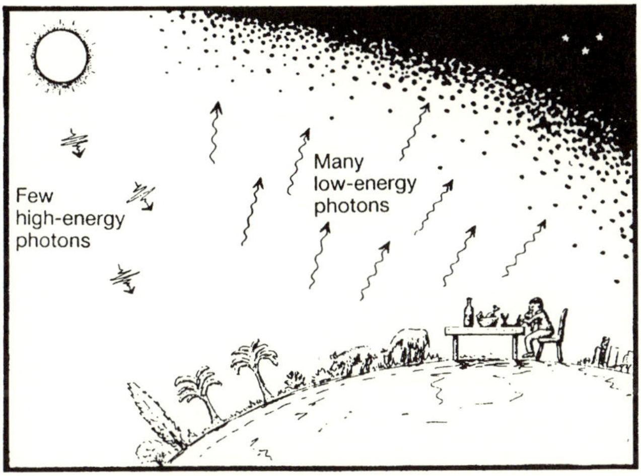

Figure 12. Artist's illustration of a fundamental reality. Living beings develop on the Earth by taking advantage of the flow of entropy between the energetic and relatively few photons emitted by the hot Sun and the enormous quantity of low-energy infrared photons that the Earth returns to space. (*Source:* R. Penrose, *The Emperor's New Mind* [Oxford: Oxford University Press, 1989], 319.)

that the positions and velocities of the molecules in the initial situation are distributed purely at random, with the general hypothesis of the two-by-two independence of these parameters that this admission supposes. From then on, the congenital link between increasing entropy and initial chance was clearly drawn.[7]

Boltzmann's conclusions may be summarized in this way. If one finds oneself in the presence of a system whose initial state does not correspond to thermodynamic equilibrium, there is an overwhelming probability that during its evolution this system will approach the state of equilibrium, simply because the probability associated with that state is incomparably larger than any other. But the state of equilibrium is also, from the point of view of thermodynamics, the state of maximal entropy. There is thus a direct correlation between a system's entropy and the a priori probability of the corresponding macroscopic state. This particular epistemological status of entropy is expressed by the famous law $S = k \log W$, engraved on Boltzmann's tomb in Vienna, which links "macroscopic" entropy S to the number of "complexions"—in other words, the possible configurations in space and in states of motion—of the system's microscopic constituents. As a consequence of this equivalence between entropy and probability, the possibility of a decrease of entropy in the immediate future of a given concrete situation cannot be excluded—it is only very improbable.

. . . AND TO DISORDER

From the teachings of Boltzmann and his disciples, we usually retain entropy as a measure of the internal disorder of systems. This idea was taken up again and developed mostly after World War II. More precisely, in an isolated system composed of many elements (such as a volume of gas, which is composed of molecules), we suppose that the components tend to distribute themselves uniformly in space and that, in the same way, the total energy of the system tends to distribute itself evenly among them. This irrepressible tendency to disorder would thus form the basis of the law of increasing entropy. Classic examples include the pile of sand that, bit by bit, levels itself or a footprint on the shore, which will always end up being erased.

Some idealistic thinkers, such as Bergson, noting the link between entropy and chance (or entropy and probability), used it to argue that entropy deserves a subjective status. We can—indeed, we must—resist this temptation. We must not forget that entropy is first linked to heat exchanges at the interiors of systems and between them. It is only secondarily, and because heat itself is linked to a fundamental chance (a chance that we define as that of the aleatory agitation of molecules), that entropy is related to chance. The thermal agitation of molecules obviously has nothing to do with a subjective definition of probabilities.

To avoid any misunderstanding on the relationship between entropy and disorder, we must first understand what we mean by *disorder.* As did Peter T. Lansberg,[8] I propose the following definition for this notion. *Disorder is the ratio (D) between a system's true entropy (S) and the maximal entropy (S_{max}) to which that system might aspire, taking into account the available phase space.*

One of the great advantages of this definition is that, since it is the ratio between two entropies, it obviously has the same degree of objectivity as entropy itself. In addition, it characterizes disorder as an intensive quantity (the combination of two systems with the same degree of disorder has the same degree of disorder as each of those systems). Note as well that this definition implies that an ordered system ($D < 1$) is not a system in thermodynamic equilibrium ($S = S_{max}$).

This definition might be objected to as being obviously too formal. The notion of order seems to include a simple and general principle, the principle of economy. Remember the celebrated example of the desk or library of a professor whose degree of disorder discouraged all the new secretaries looking for work. When the professor in question was asked for a document, however, he could find it immediately in the middle of whatever pile of papers it had been stored. This was economical in terms of time and of energy for the professor, whose personal system differed from that of a trained secretary. But this example does not prove that order and disorder cannot be given objective definitions. There is order in the world, whatever the situation being considered, whenever it is possible to construct a formula condensing the description of that situation. A crystal is ordered because its structure can be defined by describing the type of regular geometric arrangement of its atoms. The laws of science are just so many examples of such economical formulas. They define an order hidden within nature. The physicist who uses them to construct a machine creates *order from order.*[9] In the example of the professor's library, the formula for the description does indeed exist, since the professor knows how to find the book he needs without searching randomly through his stacks of papers. The fact that his secretary doesn't know that formula does not change that reality.

The example that I have just discussed also shows that, to evaluate order and disorder as understood here, it is not necessary to bring up or distinguish the levels of description. It is here, it seems to me, that the notion of order is distinguished from that of information (in the subjective or intersubjective sense of that word). Certainly the professor does not catalog his books according to the positions of the atoms of which they are composed; rather, he does so as a function of their forms as objects that can be held in the hand and identified by their titles, their colors, their sizes, and so on. The formula that allows him to find his books uses symbols that are pertinent only at a certain scale of description. This "information" is possible only insofar as the system is ordered, but it is not to be confused with that order and in addition does not necessarily consume all of it. It seems to me that this is the answer that must be given to René Thom, a passionate opponent of those who believe in a scientific chance, when he invokes, in the following quotation, the relationship between order and various levels of description:

> The notion of order is fundamentally morphological, which rests in the final analysis on a geometric, spatial description of a fact. There is always, underlying such a fact, a substratum formed of interchangeable elements, whether geometric points or equally probable elementary events. . . . In this type of discourse, this reference is generally never made, since it is generally impossible to make. Even better, in the systems that present different hierarchical levels of organization, the notion of order is *relative* to a certain level of organization and could never be considered as absolute. Thus, in a molecular system, perfect disorder, absolute at the molecular level, may at the macroscopic level be considered as perfect order, since all the points of the environment then have the same observable properties.[10]

If the definition of disorder that I have given previously is adopted, disorder is indeed maximal in Thom's example, notwithstanding the apparent homogeneity of which Thom speaks. In such an example, S/S_{max}, and thus D, is obviously equal to 1. It does not even seem necessary to repeat Prigogine's retort to Thom that in the system cited, the absence of correlations between components would lead him to refuse to attribute an order to them.

By using this example, Thom meant to emphasize the reasonable idea that the *subjective* notion of order that we might have depends on the level of detail with which we demand the system under study be described. But we can see that here is to be found rather the reputedly intuitive notion of *information* (that one possesses for a system). In the article cited earlier, Landsberg also considered this type of order or of information, which he proposed calling "macroentropy." Insofar as it is a question of a subjective notion—or at least intersubjective, since it requires that the different observers agree on the adopted level of precision of their observations—I prefer to use the term *information.* It connects with information_2 (information in the second sense), which I will examine in the next chapter.

All right so far. But all this is still terribly theoretical. The true entropy of a system (and even more, its maximal entropy) is not measured; it is calculated. Can one measure the disorder of a system? We must recognize, in all accuracy, that it cannot truly be done. The notion of order, as introduced here, connects with the algorithmic notion of complexity, which will be discussed in chapter 14. The exact determination of the algorithmic complexity of an object—or in other words, the length of the shortest computer code that can describe it in exhaustive detail—is most often an "undecidable" (insoluble) logical problem, in the sense attributed to this term by mathematicians ever since Gödel. Thus, one can never be sure that the degree of order attributed to a system is in truth the maximal degree that can be attributed to it. But the fact that we can not, beyond the shadow of a doubt, determine a system's degree of order does not mean that the notion of order is not entitled to be numbered in the arsenal of our objective theoretical concepts.

Until the recent past, it was thought that the law of increasing entropy reflected the tendency of systems to adopt a state of thermodynamic equilibrium, characterized by components in the greatest possible level of disorder among all the various degrees of freedom available to them. As a consequence, the universe as a whole was thought to be progressing ineluctably toward total destructurization and disorganization. The universe's destiny seemed terrifying, a lukewarm and gray end. Life appeared to be a desperate venture in the face of the law of increasing entropy, an untiring but, in the end, vain attempt to rescale the slope, as doomed to failure as was Sisyphus. In *L'Évolution créatrice* (Creative evolution) Bergson made himself the most fervent spokesperson for this view, even though he did not share Camus's pessimistic conclusion in *Le Mythe de Sisyphe* (The myth of Sisyphus). About twenty years ago, it came to be understood (thanks to the development of the thermodynamics of irreversible processes) that the construction of complex systems does not contradict the law of increasing entropy but on the contrary relies on it. It is the mechanisms of this autocomplexification, which benefit from the virtues of chance, that I will now examine.

ORDER THROUGH NOISE

The systems encountered in nature that spontaneously adopt a complex structure, whether living or inert, must pay a price to develop and maintain that complexity. Thermodynamics bars them from meeting their needs by depleting their own resources. To achieve their complexity, these systems must be open and find what they need for growth in random external events; in other words, they must suck new complexity from the external bath of chance. This is the principle of order through noise. Living order is comparable not to the rigid geometry of crystal but rather to that of continuously renewed plumes of smoke, as Henri Atlan poetically described it.[11] But the question remains: how can noise inspire order if it is not itself the potential bearer of an organizing principle?

At this point we must get rid of an ambiguity. The creation of order from noise is not in itself paradoxical. In the following chapter, sacrificing clarity for a terminology that has unfortunately been consecrated by use, I will identify order with *information* (in the first sense). We must not lose sight of the fact that such information is defined *apart from any semantic meaning*. Thus, order created through noise is semantically neutral. There is no creation of *meaning* in these systems, and from this point of view, human beings, produced by the evolution of living systems following the principle of creation of order through noise, are necessarily as absurd as is chance itself, as Camus and Monod have clearly seen. A complete discussion of this situation, however, is not possible until after the notion of information has been better understood. It will confirm that there is indeed homogeneity and not difference in nature between information and noise.

Another characteristic of order, insofar as it contrasts with disorder, is that it presupposes the existence of correlations. Prigogine and Stengers illustrate this idea very well.

> For a long time turbulence seemed an obvious example of disorder, par excellence. In contrast, the crystal appeared the obvious model of order. . . . We are now able to modify those judgments. We know today that turbulent flow must be seen as "ordered." The motions of two molecules at a macroscopic distance from each other, a distance potentially measured in centimeters, are in fact related to each other. By contrast, the atoms that form a crystal vibrate incoherently around their points of equilibrium. The crystal is disordered from the point of view of its modes of excitation.[12]

Comparing Prigogine's and Stengers's positions with that of Thom, the extent of the *semantic* misunderstanding is striking.

NOISE AND AUTO-ORGANIZATION

Two characteristics, among others, distinguish living complex systems. On the one hand, as already noted by F. Jacob in *La Logique du vivant* (The logic of life), these systems are organized in successive, relatively autonomous stages, the "integrons": nucleus, organelle, cell, organ, system, organism. Each integron is composed of an internal network of metabolic regulatory systems. On the other hand, each integron is far from being enclosed, closed in on itself. It is, in a way essential to its development and to its survival, open toward the exterior, with which it exchanges matter, energy, and entropy. Several authors (Henri Atlan, Edgar Morin, Francisco Varela) were struck by the observation that noise is necessary for complex living systems to develop their own complexity.

In *Entre le cristal et la fumée* (Between crystal and smoke), Atlan explains that in a complex system with several levels of integration, the transmission of information between two similar parts of a single system can induce errors that, from the point of view of the transmission of information, constitute a loss of the message's fidelity. From the point of view of the integrated organism, however, this loss implies a lessening of the redundancy linked to the repetition of units at each level of the integron and therefore an increase in the organism's complexity and degree of organization. He summarizes his astonishment at the knowledge that life develops and nourishes itself by taking advantage of errors with these words:

> But what are these errors? From what we have just seen, due to their positive effects, they no longer quite seem to be errors. The noise the aleatory factors of the environment cause in the system is no longer truly noise from the moment in which it is used by the system as an organizing factor. This would mean that environmental factors are not aleatory. But they are. Or more exactly, whether they are recognized a posteriori as being either aleatory or part of an organization depends on the system's subsequent reaction to them. A priori they are in fact aleatory, if one defines chance as the intersection of two independent chains of causality. The causes of their unforeseen intersection have nothing to do with the sequence of events that constituted the history of the system until that time.[13]

Strange power of chance! From chance order finds sustenance. Could it be that the notion of disorder might be richer than that of chance, as Morin suggests?[14] Or could it be that chance itself, not as we imagine it, leashed by our theories, but as it is encountered in nature, might be richer than we have ever thought?

ENTROPY AND INFORMATION

11

To speak of equivalence between entropy and information leads to one of the greatest misunderstandings of the history of cybernetics. Many authors have called attention to this incongruity.[1] In 1940 Claude Shannon, then a twenty-four-year-old researcher at Princeton University, cleaned up the principles of information theory. He was not initially happy with the idea of using the term *entropy* in this context. He thought that "uncertainty of a message" would be less ambiguous. It was, reports Jeremy Campbell,[2] John Von Neumann, one of the most important founders of cybernetics, who persuaded him to adopt the word *entropy* by saying: "Since no one understands what entropy is, by using this word you will have an advantage over your adversary in any debate!" This mistrust concerning the concept of entropy is surprising coming from such a rigorous mathematician and physicist. I think, along with many others, that it is possible and necessary to give entropy a purely objective status by limiting its use to that of a thermodynamic context.[3] Of course, this presupposes that we agree to attribute an objective status to heat, thus to molecular disorder, and thus to chance. They all go together.

In the pursuit of Boltzmann's original idea on the relationship between entropy and disorder, entropy came to be seen as a measure of the distance between our knowledge of a system's state and its true state and thus a reflection of the rough and approximate character of our measurements. According to this idea, just as the entropy of a system grows when we follow its evolution over time, so do the details of our knowledge of its true dynamic state become blurred and so does our uncertainty about it increase. We can show that the increase of a macroscopic system's entropy cannot be linked to the loss of information about the true state of that system over time unless we admit our inability to determine, from the beginning, the exact state of that system. Of course, in practice things do indeed happen in this way. In all but rare cases, the state of a system can never be determined with absolute certainty. Moreover, because of the natural evolution of the system, this initial uncertainty about its state becomes greater over time unless a new measurement is taken. On the other hand, a perfect determination of the initial state would allow a perfect prediction of the system's state at every moment in its future, with no loss of information. To justify our inability to determine the exact state of a system, some insist that our instruments of measurement are too crude to allow for a perfect measurement, one beyond all uncertainties. Others say that in any event, our minds would be incapable of grasping and handling the enormous quantity of details that would become accessible with more and more powerful instruments. The equivalence between the increase of entropy and the loss of information is the principal result obtained by these so-called macroscopic or coarse-graining approaches to the information available on a system. In a world where the detail of the information, the perfection of the measurements, would not be limited, the acquired information would not be lost. We can even show (this is the aim of the fundamental theorem of statistical thermodynamics, called the "Liouville theorem") that the information would be strictly preserved over the course of a system's natural evolution.

What remains is that the entropy of physical systems effectively follows an inexorably progressive course, parallel to the loss of coarse-grained information. One way to understand this parallelism would be to give a purely subjective interpretation to the concept of entropy, as proposed by many theoreticians of statistical mechanics. I will not follow them down this pathway, which is dangerously open to a syncretism between physics and metaphysics, even if I am ready to admit that in certain circumstances the entropic balance sheet must, as Léon Brillouin—the French successor of Shannon and Von Neumann—wanted, take account of the reserves of negative entropy existing in the form of information (discussed later). Entropy, the thermodynamic quantity linked to heat exchanges, derives its existence from the chance agitation of the atoms, not from the mental principles of order or of subjective information individual to each observer. As so strongly suspected by Sir Arthur Eddington, the increasing entropy of any given system, even one as small as a Pascalian *ciron,* is apparently linked to the inevitable interaction of that system with the external world.[4] For example, external influences might impose a certain distribution of the internal degrees of freedom of the system's most minute constituents (like Boltzmann's molecular chaos), which would from then on cause a growth of entropy. The theory covering such mechanisms, needless to say, has not yet been completely developed.

THE TWO TYPES OF CYBERNETIC INFORMATION

The debates over how to measure information in physical systems testify at the very least to great confusion. Just as with the notion of order taken in its subjective sense, the notion of information often presupposes that we are dealing with a system having relatively autonomous levels of description. Happily, nature abounds in examples of this type. It displays for us the levels of the "elementary" particles, of the atoms, of molecules, of objects enclosed by an external surface. The number of possible levels of description of living systems is no less impressive. There are DNA and other chemical macromolecules, the cellular organelles, the cell, the organ, and the individual.

One not insignificant part of the confusion arises from the conventions adopted by Shannon and Brillouin in their pioneering works, which identified information with entropy (both being evaluated with the formula $S = -k\Sigma p \log p$, with Brillouin using the inverse sign for information).

As Jean Tonnelat noted, however, "Contrary to what has been written, . . . a system does not know its own state and does not possess 'information' that would be present but unexpressed."[5] In contrast, a system can exhibit in its internal structure a type of order such as that evaluated in the preceding chapter. Such a statement remains true at the biological level, and it is only through an abuse of language that we can speak of "the information contained in the genetic code," for example.

Since history has nevertheless consecrated the use of the word *information* in this context, we must at least distinguish two categories of "information" while taking care to specify that in cybernetics no hint of "meaning" (semantic) is concerned.

1. When speaking of the information contained in a message, in a nucleic acid, or in an object of any kind, we are speaking of the information that would be necessary to describe the system without ambiguity (or to reconstruct it from those elements, as with building blocks). Such a definition requires that one specify, at the beginning, the scale on which the description is based (atoms, nucleotides, etc.), or in other words, what alphabet of elements we plan to use to reconstruct it. I will call this definition of the word *information* "the second sense," abbreviated as "information_2." We can also speak of "missing information," of "relative information," or even simply of information if we abstain from using this term in other contexts.

2. Another type of information is to be found more and more often in the literature—information that does not require the scale of the description to be defined and that is a purely objective magnitude. It is closely related, in fact, to the notion of order considered in the preceding chapter. It could be called, through a clear abuse of language, the "absolute information" of the system, or "information_1." The use of the concept of order or, above all, of "complexity" seems to me to be much better in this context. But if we retain the use of the term *information,* we must in any case distinguish this usage from the preceding one.

Thus, David Layzer writes: "I define the information **I** [my information$_1$] associated with a statistical description by the formula: $\mathbf{I} = H_{max} - H$, where H is entropy and H_{max} is the maximal value of H subject to given constraints."[6] And he adds: "Some authors (e.g., Brillouin) define information simply as negentropy $-H$. The difference between this definition and the present one is important because H_{max} is not in general constant during information-generating processes. It increases or decreases when the number of states accessible to a given system increases or decreases."

It is this definition, for example, that Hubert Reeves uses when he describes the information created during the cosmic nucleosynthesis phase, that period close to the Big Bang during which the atomic nuclei were formed by assembly or "fusion" of elementary particles. The most stable of all the atomic nuclei is iron, and yet during that phase the vast majority of matter was created not in the form of this metal but in the form of light elements such as hydrogen and helium. The universe is ipso facto not in its state of maximal entropy.

"The nuclear reaction that created the helium nuclei, during the first minutes, emitted a certain number of photons into space," he writes.[7] "Absorbed and reemitted by the ambient matter, these photons effectively increased cosmic entropy. But they increased it much less than it would have been increased if *all* matter had been transformed into iron." Continuing on this subject, Reeves comments on the notion of "delay," of the time lag between the appearance of an amount of negentropy and its dissipation. The information that appeared in the nucleosynthesis, he explains, actually measures the *difference* between the maximal entropy possible and that which was actually realized by the forces of nucleosynthesis at any particular moment in the history of the Big Bang.

"Like nuclear entropy," he continues, "nuclear information is *null* before the first minute. It appears because of the inability of the nuclear force to keep up with the too rapid rate of the expansion and to give free reign to its powers of synthesis. It is held back by all the nuclei that *are not* iron nuclei."

The idea of "delay," of a time lag between maximal and realized entropy, was vigorously exploited by Brooks and Wiley in their presentation of evolution as a purely entropic process, insofar as the complexification of organisms (through acquisition of complexity by way of the order-from-noise principle) allows H_{max} to grow much more quickly than entropy. In these conditions, even though entropy is necessarily growing in conformity with the second principle of thermodynamics, the information of living systems—sometimes also called "negentropy"—grows also. But these authors distinguish—unduly, I think—two parallel entropic phenomena: the evolution of the biological information specifying the historical pathways of ecological life, on the one hand, and the degradation of energy in the energy flow that traverses physical or biophysical systems, on the other hand.[8]

In fact, contrary to their assertions, it is not necessary to distinguish two laws of irreversibility, one having to do with the dilution of information and the other with the dissipation of heat. The only root of irreversibility is of thermodynamic origin. It is physical and not mental. Of course, in a second stage the mental processes, or those that call memory into play, can add a new dimension to this irreversibility, but it is not necessary to call on these processes at this stage to justify irreversibility.

MENTAL INFORMATION IS AN "ENTROPY SAVER"

These remarks are not aimed at denying any relationship between entropy and information, even if the information is subjective—that is, acquired by a thinking creature. An informed person is a person able to act, and he or she can use this ability locally to diminish increases of entropy. Thus, when the effort is made to learn a text by heart, it can later be reproduced and re-created in that way in ordered systems of letters and words, even if the original is damaged or destroyed (Michel Serres used to say, charmingly, that each poem learned is a victory over death). In addition, our faculty for organization allows us, through will alone it seems, to create order where disorder reigns. Memory and will thus seem to be a means of thwarting the law of increasing entropy. By studying the relationships between information$_2$ (measured in bits—that is, by the number of binary digits needed to code it) and entropy, Brillouin in 1948 reached the interesting conclusion that these two magnitudes are indeed convertible into each other (to a proportional constant ensuring the homogeneity of the variables), but that for a closed system the sum information$_2$ + entropy (with the appropriate conventions of signs) will inevitably continue to increase.

Information$_2$ is thus a means of postponing an increase of entropy, of provisionally improving the balance sheet without being able to reverse its inevitable course. Will does indeed allow matter to be organized by using previously acquired information$_2$, but constructing this information$_2$ requires spending more in the way of entropy than the amount of order that is finally created.

One of the contemporary epistemologists who has given the most thought to the relationship between entropy and information in this sense is Olivier Costa de Beauregard. As he very rightly stressed,[9] the remarkable fact in all this is the extraordinarily smallness (relative to our usual standards) of the transformation constant of changing information$_2$ into reduction of entropy (or negentropy): k log 2 approaches 10^{-23} J deg^{-1} bit^{-1}. In others words, 10^{23} bits of information (the equivalent of several billions of billions of pages of text) are needed to counterbalance a single Joule per degree of entropy (the entropy brought on when heating a one-liter pan of water from room temperature by a little less than a tenth of a degree). It follows that acquiring $information_2$ is very cheap in terms of negentropy, but producing negentropy (organizing matter) is very expensive in terms of $information_2$. All our social lives, states Costa de Beauregard, are colored by this simple fact: any physical measurement, even if it teaches us a great deal about a system, increases the entropy of the environment by only a little bit;[10] this is why we so carelessly accumulate volumes of information. On the other hand, diminishing the entropy of a system by making it more organized than it was requires quantities of information that rapidly become phenomenal (this is what Maxwell's demon strives to do, without ever accomplishing it). We do not care to destroy negentropy in order to diffuse $information_2$ in the form of brochures that many recipients immediately throw away. On the other hand, Costa de Beauregard again notes, one of the principal ways of reducing the cost of mass production is to reduce the human labor involved, since willful intervention always costs a lot in energy and entropy.

"Thus," states that author,

> the very smallness of the constant k, or k log 2, in practical units, expresses an existential situation where observation is easy and action difficult: the situation irreversible in fact, but reversible by right, described by Carnot's principle, where causality prevails over finality in the end.
>
> What would happen if we let $k \rightarrow 0$? This describes a ludicrous situation where observation is completely free and action absolutely impossible, a theory that had its moment of glory under the name of "the theory of consciousness as epiphenomenon." Cybernetics, by contrast, made the important discovery that *one can't get something for nothing, not even an observation* (Gabor). It made consciousness pay the price of a spectator—a very small sum—but at the same time it freed it and allowed it to be an actor also.[11]

To summarize, remember this. An increase of entropy can be interpreted as a gain of $information_2$ *that we would obtain by measuring with all the necessary precision the real state of the system* (expressed in terms of the alphabet used to describe the system at the level considered). The use of a statement relating to an experiment that we could in principle conduct but that we do not really need to conduct allows us to avoid having to know how, in concrete terms, we would proceed to obtain such information, which thus remains in its potential state. This is unrealized information.

It is easy to show, in statistical thermodynamics, that the heat capacities of monatomic gases (their specific heat capacities at a constant temperature or volume, which can be measured experimentally) have exactly the value predicted by the Shannon-Brillouin theory, such that increases in volume or temperature correspond to the loss of information brought about by the corresponding processes: the isothermic expansion of the gas increases uncertainty about the position of the individual gas atoms, and heating at a constant volume widens the dispersion of their velocities. Another example is furnished by the decay of a radioactive substance. I have already had the opportunity to mention elsewhere[12] that at least one part of the increase of entropy observed during this phenomenon can easily be understood by noting that this transformation is accompanied by the appearance, at the heart of the initial homogeneous sample, of a new type of atom, which is the residue of the decayed atoms. The counters usually used in this type of experiment tell us how many atoms decay in each unit of time, but they do not tell us which ones decay. Therefore, at the heart of the radioactive substance there appears a new ambiguity concerning the nature—decayed or not—of each particular atom of the initial sample. The missing information that results corresponds, according to Shannon's theory, to a minimal increase of the thermodynamic entropy in this process.

Henri Atlan contributed one of the most complete and pertinent studies[13] on the qualitative and quantitative usage of the notion of information$_2$ in biology. He concluded that every time we speak of information, it is information "that an observer must have" to describe a situation or reconstruct a system from its component parts. This clarification removes the ambiguity about the status of the information contained, for example, in a DNA molecule and its relationship to entropy.

INFORMATION DOES NOT GET LOST, BUT IT DOES CHANGE SCALE

In contrast to macroscopic or relative information, information in the first sense does not get lost, since the future of a state that is perfectly identified is perfectly determined.

To better understand this fact and its relationship to the loss of information in the second sense, we must examine some of the details of the demonstration, but I will keep that examination as nontechnical as possible.

During the natural evolution of a physical system, the amount of "microscopic" information necessary to describe its state remains constant over time. This is in fact implicitly contained, as already mentioned, in Liouville's theorem. It is true, however, that in many concrete situations the *quality* of the information changes. The information needed to specify the initial state usually is rather of a *macroscopic* character (for example, the fact that all the molecules of a drop of ink that will be added to a glass of water must be gathered in the same small volume v enormously simplifies the macroscopic description of the situation), while that which is necessary to specify the final state is *microscopic* (to describe the situation of the ink after diffusion in the water contained in the glass, we cannot escape a fastidious enumeration of the position of each individual molecule). Briefly, an absolutely precise description of the initial situation is not only generally impossible but also entirely useless in practice. In this case the rough description of the initial state does not allow a comparably complete description of the final state. Uncertainties always tend to increase.

The first element that is essential to account for irreversibility is a statistical supposition about the initial state of a macroscopic system. In practice this consists of saying that all the microscopic states compatible with the given macroscopic conditions (temperature, pressure, etc.) are equally probable. More concisely it means that in their initial states, at the moment at which they are isolated for a study of their behavior, natural systems do not possess microinformation. For example, very detailed information about the correlations existing between the molecules of a perfect gas are useless in characterizing the initial state—as useless as the knowledge, in quantum mechanics, of the relative phase of the wave function representative of each of the systems of an "ensemble."

Next the reasoning uses the two following arguments. The first confirms that the global information does not change, and the second shows that it changes scale, going from the macroscopic to the microscopic level.

1. Since the precise state of the system is not known, a probability p_k is associated with each microscopic state k. This is the *principle of microscopic uncertainty.* This principle is necessary but not sufficient, since Liouville's theorem says that probability p_k does not change over time. In other words, it is always possible to associate unequivocally each possible final state to an equally possible initial state. Therefore formal entropy $H = \Sigma_k p_k \log p_k$ remains constant over time.

2. In the second stage, we use the *principle of macroscopic uncertainty.* Such a principle was proposed by Willard Gibbs under the term *coarse graining,* which is now classic in the circles of thermodynamic statisticians. In this procedure, each macrostate a is in fact a group of microstates i that we declare ourselves incapable of distinguishing from one another. The probabilities p_a are given by the sum of the probabilities p_i of the corresponding microstates belonging to the macrostate a. In this case, simple reasoning allows us to write the information in the form of two terms, one ($\bar{H}$) relative to the macroscopic information and the other (H') relative to the microscopic information.[14]

Since the total information remains constant, $\bar{H}$ and H' must vary inversely. And because, as we have already seen, the growth of thermodynamic entropy parallels the loss of macroscopic information, the microscopic information must be increasing. Entropy must therefore "flow" from the macroscopic degrees of freedom toward the microscopic degrees of freedom. This situation was admirably summarized by Émile Borel in 1914: "The evolution of the universe could thus be pictured as tending to produce a more and more complicated state, not perceptible or usable except by smaller and smaller beings."[15]

Recently the relationship between entropy and information has been at the heart of some new works. A series of articles by W. Zurek, of the Los Alamos laboratory, confirmed the preceding findings using a new definition of entropy that introduces a form of Gödel's incompleteness theorem into physics. Zurek's works explicitly allow the introduction of the observer and the information he or she acquires during the act of measurement.[16]

CONCEIVING AN OBJECTIVE SOURCE OF INFORMATION

The increase of entropy as described by classical thermodynamics is thus reached only after two stages.[17] In the first stage, information changes scale without being actually lost. At this stage the increase of entropy is only apparent or subjective. It appears only insofar as we introduce into the theory limitations to the detailed descriptions of systems, limitations linked to the capabilities (which are themselves limited) of our investigative tools.[18] But in a second stage, this increase of subjective entropy acquires an objective, ontological status. Here we must imagine that there exists a rupture in the chain of information: a substratum in which correlations cease spreading and becoming more and more subtle and are purely and simply erased. If such a break exists, information could dilute itself irreversibly in this "sink of correlations." This ugly expression, created by Thomas Gold,[19] lacks poetry. I might prefer to invoke the universal "great chaos," if the term *chaos* had not already assumed a particularly ambiguous meaning over these past few years. By using the word *chaos*, I do not mean the deterministic chaos wooed by modern mathematicians (which is the realm not of chance but only of its appearances); rather, I mean Plato's original Chaos, the universal substratum that preceded raw matter and on which, according to that philosopher, the demiurge exerted his power of organization, a notion that found its natural extension in Boltzmann's molecular chaos.

The idea of linking the growth of entropy in phenomena at our scale to the expansion of the universe seems to clash with the statement that this expansion is overall an adiabatic process[20] and thus does not entail an increase of entropy. But the fact that the expansion of the universe does not cause an overall increase in entropy is not a definitive argument against the suggested linkage. There are, from all the evidence, local phenomena of increasing entropy, such as the aggregation of the stars and galaxies, which obviously depend on the dynamics of universal expansion. In addition, the suggested linkage is certainly to be sought not in the mechanisms of entropy growth at various scales that we could consider but rather in the diffusion of information from the macroscopic toward the microscopic scale that I have just considered. The growth of entropy will always rely on the question of the structure of the universal substratum, which is the true "sink of information." Could this not be a property we might be tempted to attribute to ontological chance, whose traces we are seeking?

Today it is incontestably the American astronomer D. Layzer who has returned most consistently and systematically, and without being too bothered by the skepticism of his colleagues,[21] to this idea of a break in the ladder of information. He discussed it in particular in his book *Cosmogenesis.* I will come back to his ideas in chapter 13, but first we must take a look together at the new manifestations of natural chance, this time in microscopic physics.

QUANTUM INDETERMINISM: THE GAP BETWEEN THE REAL IN ITSELF AND THE REAL IN PRACTICE

12

CHANCE DUE TO IGNORANCE OR ESSENTIAL CHANCE?

What are the origins and the true significance of chance in statistical theories? What limits does the use of instruments impose on our understanding of the true states of systems? Boltzmann, Maxwell, Gibbs, and the other founders of classical statistical theory had to confront these questions well before the invention of quantum mechanics. These same questions reappear now in quantum theory.

However, the difference introduced by quantum mechanics is enormous. The fate of a classical system in unstable equilibrium is always resolved by some influence, even if that influence is only a fluctuation of temperature and even if the reason escapes the experimenter. A quantum system resolves itself without a reason, or rather, because of an internal "propensity" that does not follow any determination. This is at least the position of the vast majority of physicists who live with the daily routine of the laboratory. The metaphysics that underlies the "orthodox" interpretation of quantum mechanics evokes, twenty centuries later, Lucretius's *clinamen:* fundamental chance resides in the atoms themselves. That is a statement that cannot help but shock many deterministic rationalists. But how can we not marvel at quantum theory's fantastic power of prediction, despite such a strange foundation? The theory of quantum probabilities not only unseats determinism; it brings even greater ruination to reductionism. In so doing, it continues and expands the movement begun in the nineteenth century by Maxwell.[1] This is why some thinkers, such as René Thom, nostalgic for Laplacian enthusiasm, resist stubbornly and will under no terms surrender their resistance.

In truth, Lucretius's heritage in the invention of quantum chance must be strongly qualified. Quantum mechanics requires the claim that fundamental chance resides not in the atoms *themselves* but rather *in the atoms as they mani-*

fest themselves to us. The quantum object is neither a wave nor a particle whose true state is unknown but rather something that has no properties (except perhaps the property of being) before the type of investigation to which it is to be subjected is defined. Quantities, such as position, momentum, or components of angular moment, cannot be applied to these objects except in concrete situations. This is at least the basic lesson taught by Niels Bohr in 1927 and today accepted by the vast majority of physicists.

THE QUANTUM PRINCIPLE OF SUPERPOSITION: A CHALLENGE TO CLASSICAL THINKING

The general impossibility of attributing a definite value to each quantity that might characterize a quantum object comes from the principle of superposition. Thus, the position of a particle, for example, may not only take any value within the limits assigned by the walls of the apparatus in which it is enclosed, but the particle itself may exist in a *superposition of states,* each corresponding to a given position. The "ket," which in this theory represents the particle, will be the sum of the kets of its various possible positions. The same may be true for the particle's momentum, its angular momentum, and so on. A classic example of the concrete manifestations of this possibility of superposition is furnished by polarized light. Polarized light leaving a linear polarizer (a polarizing filter) is in a state of equal superposition of left and right circular polarizations. Inversely, circularly polarized light emitted by a given atomic transition is in a state of equal superposition of the linear polarizations defined by two directions forming a right angle between them and perpendicular to the direction of the light's propagation.

To prove that the objects studied in our laboratories conform to this possibility—in other words, that the individual quantum objects are indeed generally in a state of superposition (and that we are not simply dealing with statistical mixtures of objects, each with a well-defined state)—we must observe the characteristic forms of interference effects between the different components of the ket representing a single system. Such phenomena (diffraction rings or fringes of interference) are currently observed in optics when different components of a single wave combine their effects at a single point.

Interference patterns produced with the aid of electrons were observed for the first time between 1921–23 by Clinton Davisson and Charles Kunsman at the Bell Laboratories of New York, but they were not recognized for what they were at that time. In 1925 Walter Elsasser, a young physicist who had read de Broglie's report on the possibility of waves associated with electrons,[2] suggested interpreting them as phenomena of coherent superposition (diffraction). When he informed Einstein of this interpretation, the former exclaimed: "Young man, you are sitting atop a gold mine!"[3] The experiments of 1921–23 were not clear enough to be interpreted without dispute, but some years later Davisson and his colleague Lester Germer, who were studying the reflection of an electron beam on the surface of a crystal, obtained images very similar to the forms of reflection on a diffraction grating obtained with light waves, irrefutably proving the existence of the property of superposition in the case of electrons.[4] In the United Kingdom, George Thompson obtained superb images of diffraction rings with electron beams passing through thin sheets of metal.[5]

Can the principle of superposition be rejected? To do so, we would have to imagine that the ket describing a quantum system, if it does indeed describe what we know of the system, does not represent all that we might know about it. This in fact was Einstein's position. He developed arguments in this direction in the 1935 article he signed with Boris Podolsky and Nathan Rosen.[6] Then an attempt was made to develop theories "with hidden variables," designed to complete the description in terms of kets. Thanks to experiments on inseparability, we know today that any eventual hidden variable can never be other than nonlocal (its value would certainly not be determined solely by what is happening in the immediate vicinity of the point being considered but would depend on distant "influences"). Because of this complication, this line of reasoning is largely ignored today except by a few researchers.[7]

Moreover, recent detailed experiments have directly proved that a single photon can be simultaneously in two distinct branches of an interferometer.[8] It seems thus that the principle of superposition is based on solid experimental findings. But this principle leads quite naturally to a recognition that chance has a very special status in quantum mechanics. We are indeed talking about an essential chance and not a chance of ignorance, because the photon, which when diffracting through a diaphragm is described by a superposition of waves, is in no way constrained to appear at any one particular spot on the screen rather than at another. In the same way, the unstable particle that is described by a superposition of nondecayed and decayed states is in no way constrained to manifest that radioactive transition at one particular moment rather than at another.

DETERMINISM AND THE "COLLAPSE OF THE WAVE PACKET"

Historically the most difficult battle the founders of quantum mechanics had to fight in the thirties was with determinism. The indeterminism in this theory lies in, for example, the absolute impossibility of which I have just spoken: predicting the precise moment at which a particular radioactive particle will decay. That indeterminism depends on another axiom of the theory, the principle of collapse of the wave packet. Even today the epistemological status of this notion is the subject of passionate debate, since it seems difficult to reconcile with the determinism of classical mechanics. How can macroscopic objects obey classical determinism when they are, in the end, made up of microscopic objects that do *not* obey it?

Several avenues were open to theoreticians attempting to surmount this difficulty. Perhaps the formulation of quantum mechanics in its current form is only a preliminary sketch that should be modified, and the theory that will replace it will reconcile classical determinism with the new theory describing microscopic systems. That was Albert Einstein's hope. Perhaps it is classical mechanics that must be consigned to the pigeonhole reserved for excessive simplifications, as a pretty theory in principle but inapplicable in practice, inasmuch as concrete systems will always retain the trace of impishness, if we dare express it thus, necessary to outlaw any strictly deterministic prediction. This is the thesis defended by I. Prigogine and I. Stengers.[9]

Debates on the true status of the principle of collapse of wave packets excited the founders of quantum mechanics from the beginning, as the excellent selection of articles assembled by J. A. Wheeler and W. H. Zurek reminds us.[10] In his recent works, d'Espagnat demonstrates that, if several attitudes are possible, the answers that most directly accord with both pure materialism and pure realism are impossible to reconcile with our current knowledge (this judgment applies, especially, to the solution Einstein hoped to construct).[11]

Why did such great thinkers try so hard to restore determinism to physical theory? Obviously there is something metaphysical in this desire to hold on to determinism. It is an attempt, finally, to save humankind from the absurd, to give us at least the power to be able to know. How can one set about constructing a science of the real if the ideal of determinism is abandoned? In his monumental book more than thirty years ago, Popper had already commented on this idea.

> The belief in causality is metaphysical. . . . For a statement supposed to be free from metaphysical elements can have indeterminist conclusions only if they are falsifiable. But they can be shown to be false only if we succeed in formulating laws, and in deducing predictions from them which are corroborated. Accordingly, if we assume that these indeterminist conclusions are *empirical hypotheses,* we ought to try hard to test them, i.e., to falsify them. And this means that we ought to *search* for laws and predictions. Thus, we cannot obey an exhortation to abandon this search without repudiating the empirical character of these hypotheses. This shows that it would be self-contradictory to think that any empirical hypothesis could exist which might compel us to abandon the search for laws.[12]

We can only regret that in this text Popper used the term of *causality* where it seems *determinism* would have been better. Popper was of course up to date on the developments of quantum mechanics, and he explicitly recognized that the formulas of quantum mechanics do not necessarily lead to acausal conclusions, because although the teachings of quantum mechanics and the experiments it suggests oblige us to keep our distance from Laplacian determinism, they do so without demanding that we renounce all causal paradigms. It is precisely this, in my opinion, that makes quantum mechanics special in this respect.

As a matter of fact, the evolution of the representative ket of a quantum mechanical system is governed by a differential equation of the first order relative to time (Schrödinger's equation; see appendix 3) entirely comparable to Newton's equation in classical mechanics. From this point of view, it is indisputably a causal theory. On the other hand, what evolves over time is the ket—that is, the group of "propensities" relative to the possible results of measurement—and not the physical quantities associated in classical mechanics with the system itself. Thus quantum mechanics is indeed a causal theory, but it is not deterministic.[13]

SEPARATING THE MICROPHYSICAL WORLD FROM THE PHYSICAL WORLD

When he participated in the creation of quantum mechanics, Werner Heisenberg was at first tempted to research the hidden parameters that would permit the restoration of absolute determinism to the description of microscopic phenomena. But he soon decided to oppose that tack, deeming it to be hopeless. The reason he gave concerned the separation between the observed object and the apparatus of measurement, a separation that was believed to delineate the separation between the areas of validity of the quantum description and that of the classical description. This barrier, he said, is fundamentally mobile. Since from each side events are causally determined (the former by Schrödinger's equation and the latter by classical mechanics), the statistical character of the predictions of the quantum theory is linked precisely, he thought, to the placement of that separation. Any appeal to additional causes (hidden parameters) in quantum mechanics would lead to contradictions between two observers who choose to place the barrier differently. This argument still seems valid today. However, many physicists continue to seek objective criteria that would allow the site of the separation between microcosm and macrocosm to be indisputably fixed. Some, with John Archibald Wheeler, propose placing that rift precisely where the thermodynamic arrow of time makes its appearance. For them, a classical system is a system endowed with an indisputable arrow of time, whereas a quantum microscopic system is not marked by time's arrow.[14]

DETERMINISM IS NOT A SINE QUA NON OF REALISM

Among the reasons physicists hate the idea of abandoning determinism is the fear that this abandonment might also entail abandoning physical realism—in other words, abandoning the idea that the theory describes, at least in part, "that which is." This fear is without foundation. For example, Prigogine and Stengers write, "There is no reason to identify the 'dream' of a return to realism with that of a return to determinism."[15] And they quote Popper: "My own point of view is that *indeterminism is compatible with realism,* and that acceptance of this fact allows us to adopt an objective, coherent epistemology, an objective interpretation of the whole of the quantum theory, and an objective interpretation of probability."[16] Bernard d'Espagnat explains that the relationship between determinism and realism is not a two-way street. "If it is true that we have trouble imagining a deterministic theory that would not be 'realistic,' . . . the inverse, on the contrary, is easily conceived. I can imagine a world in which, on the one hand, localized things would not only be 'objects of a subject' but would truly be 'objects in and of themselves,' and in which, on the other hand, a pencil (in this sense an object) balanced on its point would fall in a direction determined only by 'true chance' (in a direction which, in other words, would not be determined in advance by any parameter, known or unknown)."[17]

On the other hand, d'Espagnat thinks that the fundamental difficulties of determinism must be placed *at the level of empirical reality,* and not necessarily at that of the real in itself. In *Penser la science* he explains, "Seen by a well-informed contemporary physicist, indeterminism has become a subtle thing, since it is only at the heart of empirical reality that its existence, in all its distinct character, can be placed."[18] Here we are, then, two centuries after Kant, at the antipodes of his thinking. It is *in*determinism that would be a macroscopic illusion. Indeed, we must recognize that acts of measurement are always operations that, in the end, are designed to tell us something about the world in a language anchored in empirical and macroscopic reality. Penrose also emphasizes the fact that the act of measurement always contains an element of macroscopic amplification.[19]

THE SPECIAL ROLE OF THE ACT OF MEASURING

As demonstrated by Max Born in 1926, the introduction of probabilities into quantum mechanics unmistakably follows the introduction of complex numbers for the wave function or the ket that represents the system being studied. The kets that are superposed (added) to one another in this theory are added with "imaginary" coefficients, and the usual physical quantities cannot simultaneously all be real eigenvalues corresponding to that state. In other words, if some physical quantities attached to the system do have a well-defined value, others cannot (see appendix 3, points 2 and 4).

Similarly, the "orientation" of a ket representative of a system in the ideal space of quantum mechanics does not represent any real quantity, and its value is arbitrary. More precisely, this ket can always be multiplied by any imaginary coefficient of unitary length (we say "a pure phase," because this coefficient is necessarily in the form $e^{i\varphi}$). In a collection of identical physical systems, each one of which finds itself in a superposition of physical states, the "orientations" or relative phases attributed to each component are generally arbitrary. They are thus chosen at random. This hypothesis turns out to have properties that are as miraculous when describing the evolution of the quantum systems in statistical mechanics as they are when dealing with the hypothesis of molecular chaos, which Boltzmann proposed to describe gases in classical statistical mechanics. It is the basis of the demonstration that Wolfgang Pauli gave of the irreversibility of the evolution of quantum systems,[20] parallel to Boltzmann's classical demonstration of the spontaneous growth of entropy in a perfect gas. Basically, the two ideas are no more than transpositions of each other in quantum mechanics and classical mechanics. In both cases it is a matter of affirming that what is outside our control necessarily obeys the implacable laws of chance.

As noted by Bernard d'Espagnat, the idea of physical measurement is essential to quantum physics (as shown in appendix 3, it is a part of the very axioms of the theory). The "propensities," as Popper would say, do not become "realities" except through acts of measurement. The different possible results are distributed statistically, conforming to the probabilities dictated by the laws of quantum mechanics, according to the process called the collapse of the wave packet. This particularity is the source of many difficulties, because it introduces a solution of continuity into the evolution of systems and because it proposes a sort of duality for the law of the evolution of systems depending on whether they are actually measured. It is important to note that despite these inconveniences, it is hard to imagine how to get around the notion. The collapse of wave packets seems to be an essential ingredient in order to avoid having the wave functions dilute themselves in space, following their natural expansion over time. The collapse of wave packets also ensures that unstable elementary particles follow a law of decrease that is purely exponential over time. These are two supposed properties of reality that we will not easily disavow.[21]

Since theories of measurement that are objective in the strong sense of the term (in other words, independent of an observer finally becoming aware of the results) do not seem to exist, we might consider that contemporary physical theory is objective only in a weak sense—in other words, that it only certifies that results will be the same for any observer taking the measurement. We can also say that it is intersubjective, meaning the same thing.

QUANTUM CHANCE AND STATISTICAL CHANCE: SIMILITUDE OR OPPOSITION?

It is time to summarize what we have learned about chance from quantum mechanics. Is it a theory of essential chance, distinct from the chance of ignorance of classical probability theory? The converging arguments in favor of a positive answer to this question make me lean toward adopting it, as they do for most physicists.

One of the sources of quantum chance is to be found in Bohr's principle of complementarity: the electron is neither a wave nor a particle and thus has ambivalences that are resolved, in fact, only as a function of precise experimental conditions. But quantum indeterminism goes well beyond that. Even in the presence of a given apparatus, the electron still has possibilities, "choices of behavior," that are not determined by the measuring apparatus. For example, in an apparatus designed to measure the orientation of the magnetic moment (spin) of the electron, called the Stern and Gerlach apparatus, the electron generally has not just one single possible trajectory but two.

Does Bohr's complementarity principle induce us to adopt a positivist view, or even a prudent agnosticism as to the existence of an ultimate reality? I do not think so. In any case, it is certain that Bohr would not have subscribed to the discourse of those of his admirers who saw a mystic principle in wave-particle complementarity. He explicitly denounced them, in advance, in *Atomic Physics and Human Knowledge.* "The analogy in the purely logical character of the problems which present themselves in so widely separated fields of human interest in no way implies acceptance in atomic physics of any mysticism foreign to the true spirit of science," he wrote, explaining further that the point of view of complementarity must be considered as a logical generalization of the ideal of causality.[22]

If we believe G. Holton, Bohr found his inspiration on the principle of complementarity in the experiments of cognitive psychology demonstrating the multiplicity of the states of consciousness—which are exclusive of one another—in some hypnotized patients, experiments reported by William James in his work *The Principles of Psychology,* which Bohr liked. Insofar as this anecdote is true, it shows that Bohr, far from seeking his inspiration in dualist philosophies with a strong positivist accent, sought on the contrary to avoid metaphysics. Simply, using a personal and fruitful means of reasoning by analogy, he modeled his vision of the world on the proven complexity of the mental processes.

Far from holding any positivist interpretation of the role of chance in quantum mechanics, we perhaps must seriously reconsider the already old proposition of Léon Rosenfeld, Bohr's loyal disciple, who refused to concede that chance's status in quantum mechanics differs from its status in statistical mechanics. He argued that just as there exists a principle of position-momentum complementarity that makes the simultaneous use of these two concepts applied to a single system impossible, so in thermodynamics does the impossibility of pinpointing the initial conditions of a system arise precisely from the framework in which one places oneself, emphasizing the irreversibility of evolution to the detriment of the exact localization of that system in phase space.

This position, which Rosenfeld defended in the collective work *Louis de Broglie, physicien et penseur* (Louis de Broglie, physicist and thinker), seems to me to come close to the current thinking of Olivier Costa de Beauregard, another individual who does not see a difficulty in the choice between determinism and chance. Rather, he sees a sort of manifestation of Bohr's principle of complementarity: two irreconcilable concepts, mutually exclusive, but both valid if they are not used simultaneously.[23]

The fact that quantum mechanics has resisted both the depredations of time and philosophical attacks for more than sixty years does not mean that it will not be replaced by another theory that is more refined and all-encompassing, as all intellectual constructions in this world eventually are. But with this theory we are already approaching a sort of sanctuary of the true nature of nature, precisely through the place it accords essential chance. Quantum theory also suggests (more than it demonstrates) the idea of a strong link between temporal irreversibility and the operations of measurement. Thus, just as that theory separates the real in itself (not measured) from the real to us (that which has already been manifested to our perceptions or that which we seek to tame with our machines), it seems also to indicate that it is the operation of measurement itself that introduces irreversibility.

The separation introduced by quantum mechanics between the real itself and the real in practice, at the same time that it introduces a sort of complementarity between causality and irreversibility, draws a picture that is once again not without parallel to the classic outlines of Darwinism. Just as Darwinism posits a rupture between the microscopic world of variations and the macroscopic world of phenotypes, subject to natural selection and thus to irreversible evolution, this theory unveils a radical break in nature between the impalpable microcosm and the perceived macrocosm. Irreversibility would "profit" from the variations that the very act of measurement introduces between the "potentialities" present in the wave functions and the concretization of the measured result, just as evolution "profits" from the distance between chance variations and selection. The temporal arrows revealed in both cases are comparable. Of course, comparing is not reasoning, but the evocation of an analogy can sometimes, as the example of Niels Bohr has just reminded us, have a great heuristic value.

BEHIND THE COSMOLOGICAL RUPTURE LIES FUNDAMENTAL CHANCE

13

From whatever angle one approaches natural chance—whether from statistical mechanics, with its equal a priori probabilities; from quantum mechanics, with the problem of measurement and the collapse of wave packets; or from the biology of evolution, with its division between variation and selection—one finds a rupture at the source of the justification of the evolution of systems. This rupture not only separates chaos from order; it also and above all marks the distance between the mystery and the strangeness of the world in itself and the familiar domain of phenomena at the mercy of our scientific instruments.

Once this rupture is remarked, the question naturally arises as to which side of the rift temporality, the source and the ultimate reason for the second law of thermodynamics, should be placed. At first glance we might seriously suspect that the growth of entropy is only an illusion, linked to our condition as human observers. This being so, we evaluate and measure everything by the yardstick that is natural to us, that of meters or of millimeters, and in any case well beyond the scale of the rupture of which we are speaking. This natural scale makes it difficult for us to go beyond a certain level of precision in the description of systems and of their correlations. It incites us rather to practice rougher methods, such as that of *coarse graining*—for example, in speaking of the average position and velocity of the molecules of a volume of gas in elementary "cells" of millimetric size. However, we know that this allocating of reality into segments immediately calls for the growth of statistical entropy, whatever the segmentation adopted, as long as it is finite (see chap. 11).

But the macroscopic approach to the growth of entropy, by coarse-graining techniques, seems a little too much like sweeping the dirt under the rug. So it seems logical to postulate that after diluting themselves more and more, the correlations between the elements that compose a system truly and completely disappear. In other words, at a certain point the microscopic information is irremediably lost as a consequence of the interactions of the elements of the system with the environment. Thus, what was at the beginning perhaps only an "illusion" (the increase of entropy linked to the change in scale of the information) becomes reality when the correlations between the elements that compose a system are erased, whisked away behind the rupture as behind a curtain, when the microscopic information vanishes.[1] But where must we place this rupture beyond which the correlations objectively undo themselves? The idea that I want to examine here is that this "trap" is a property of the universe in its entirety.

It has been sixty years since we learned that the universe is expanding. The idea that temporal irreversibility is linked to cosmological expansion is natural enough, once we realize that this expansion is the only physical process marked with an obvious temporal arrow. It was notably invoked by Eddington as early as 1935. As we will see, however, it is not enough in itself to account for the second law of thermodynamics. The recourse to the hypothesis of a fundamental chance lurking, so to speak, beyond a universal rupture seems a supplemental and indispensable ingredient. The discussion of the status of this universal chance will lead us to the heart of the still passionate debates around the interpretation of quantum mechanics.

A COSMIC SOURCE OF REVERSIBILITY

In his book *Le Hasard* (1914), Émile Borel had already observed that only one centimeter of displacement of a single atom of hydrogen at the interior of the star Sirius can change the state of an ordinary volume of gas on the Earth in less than a microsecond (aside from the time needed for the propagation of the gravitational disturbance), enough to change completely the representation of that bit of gas in phase space. It is thus totally unreasonable to seek to quantify absolutely the initial conditions of the gas. In less time than it would take to accumulate all the necessary data, they would already be completely and continually transformed.

If initial conditions are perfectly known, but only for a limited portion of a system, the part of phase space accessible to this portion can only grow, and its entropy with it. We find here again the mechanism of expansion of phase space already demonstrated in the coarse-graining technique—with the difference, however, that this technique is now imposed not by the limits of our measuring instruments but by the interactions of the portion of interest with its surroundings. The idea that the universe interacts with human-scale systems to increase the latter's entropy is certainly seductive, since the entropic arrow of time does not manifest itself on the simplest systems, such as isolated atoms, but becomes more and more evident with the increasing complexity of the systems considered, whose interactions with the universe we might think to be less and less avoidable. Do not forget that Boltzmann's H theorem, in its most modern versions, has never been demonstrated except at the limit of the most vast systems conceivable: systems having an infinite number of degrees of freedom.

But the intuition of a parallelism, or even of a relationship of cause and effect, between the expansion of the universe and the arrow of time in phenomena at our scale is not enough. If it were a question of a direct and reciprocal relationship, it would seem natural that the arrow of time would similarly mark all the macroscopic systems of a given size. This is obviously not true. The rates of production of internal entropy vary greatly from one system to another. Prigogine reminds us, for example, that we cannot apply the second law to an undamped harmonic oscillator consisting of a two-body planetary system, however macroscopic and immersed it is in the expanding universe.[2]

By reason of the parallelism invoked, most physicists up until these past few years thought, as did Boltzmann, that in the past the entire universe (or at least the portion of the universe that is directly accessible to our measuring instruments) must have had a particularly low state of entropy and consequently a very high degree of organization. Only this hypothesis, it seemed, could furnish to the various systems that did and still do people this universe such an immense source of the growth of entropy. At first sight the hypothesis that the universe was initially in a state of low entropy seems difficult to reconcile with the well-known scenario of the Big Bang. As Roger Penrose remarked: "The primordial fireball was in a state of thermal equilibrium—a hot gas in expansion, but in internal thermodynamic equilibrium. Let us remember, on the other hand, that the term 'thermal equilibrium' has to do with a *maximum* state of entropy." This contradicts the second law, which "requires that in its initial state, the entropy of our Universe must be a *minimum* and not a maximum state."[3]

The apparent paradox between the supposition that the universe in its entirety, through its expansion, has constituted and continues to furnish the ultimate source of temporality and the very reasonable hypothesis that the universe was, in the very first moments of its existence, in a hot and homogeneous state of very high entropy is not difficult to overcome. While it is true that the fireball began in thermal equilibrium, the universe at that moment was very small. The fireball represented the maximal state of entropy that could be allowed for a universe of *that* small size, but that entropy was still minuscule in comparison with that which is allowed for a universe the size of the current one. During the expansion, the maximum allowed entropy increased tremendously with the size of the universe, but the true entropy of that latter remained far below this theoretical limit. The second law of thermodynamics comes from the fact that true entropy is always less than the maximal allowed entropy and in some way lags behind it.

This point of view is gaining the support of physicists of such diverse temperaments as Roger Penrose and David Layzer, for example. It is the analog of the argument that Brooks and Wiley used to explain the appearance of order during biological evolution in a process that is fundamentally entropic.

Still, considered within the framework of general relativity, the expansion of the universe occurs through a process that is comparable to the adiabatic expansion of a gas in classical thermodynamics. It is thus a globally isentropic process, that is, one without changes of entropy (it might not have been the same at the very first instants of the Big Bang, if the expansion was preceded by an inflationary phase of explosion). Other considerations can be invoked to reinforce the idea that the expansion of the universe, of and by itself, cannot be the direct and unique source of temporality. The astrophysicists and philosophers of science have often debated what would happen if, as general relativity deems possible, the universe, after its current phase of expansion, enters into a phase of contraction, preparing itself to regroup and to collapse on itself in a *Big Crunch,* symmetrical to the Big Bang. Would the arrow of time imposed by the law of increasing entropy be reversed at the instant of this change? Today the great majority of scientists agree that no such catastrophe would occur. If the expansion of the universe were to change into contraction tomorrow, the first indication of that fact would be that the red shift of galactic spectra would change to a blue shift, this change affecting first the closest galaxies (some millions of years after the event!), whereas the spectra of distant galaxies would remain shifted toward red for immense periods of time. And even this change would not keep the entropic processes of condensation of clouds of matter and so on from continuing to take place for a very long time. There is thus not an exact symmetry between expansion and contraction phases of the universe, since the history of the universe imposes its own constraints on it. The law of increasing entropy would remain valid in such a universe until the end of time (we know that the collapse of a star into a black hole is a process that *produces,* rather than *destroys,* entropy).[4]

We cannot, thus, rely only on the expansion of the universe to explain entropic temporality. Other considerations that we have not yet looked at must intervene. We already know the actor, whom we must now introduce into the play: chance.

CHANCE AND THE COSMOLOGICAL RUPTURE

In approaching this subject, I must warn the reader that we are entering into the turbulent waters of an active debate in which we are still missing the perspective necessary to take an unqualified position. The thoughts that follow will be borrowed essentially from two physicists with largely irreconcilable views of the history of the Big Bang: the American astrophysicist David Layzer and the British physicist Roger Penrose. Both have recently published works of great merit on this subject. They agree on the idea of finding a cosmological source for the growth of entropy and on the role of chance in this context. The source of chance can be specified in both cases: quantum indeterminism. But whereas Layzer proposes searching for the deus ex machina in a universal solidarity responsible for the absence of microscopic information in the universe, Penrose thinks that the responsible party resides in the still poorly elucidated properties of gravitation and in a sort of fundamental symmetry between the uncertainty of the world in which we live and the existence of black holes, devourers of information.

Layzer's thesis cannot surprise those who understand the strange solidarity that the elementary particles called "fermions," which lie at the root of the phenomenon of materiality, enjoy in quantum mechanics. This universal solidarity of fermions was discovered in 1925 by Wolfgang Pauli and is now called "Pauli's exclusion principle." It states that there cannot be, for example, two electrons that are in exactly the same quantum state. In fact, two electrons do not have independent, autonomous existence before a duly constructed and executed experiment distinguishes them: showing, for example, that one of them is in a certain laboratory here on Earth and the other on the surface of Sirius. Layzer has developed this thesis over these past few years in the framework of an extremely simplified model of the universe. This model calls on the "strong cosmological principle," which in Layzer's version can be defined thus: the universe does not contain any objective information other than the statistical. For example, it can have neither center nor preferred direction. This is fundamentally the modern version of the ancient principle earlier defined by Giordano Bruno, who paid so dearly for this attitude that was contrary to the triumphant anthropocentrism of Renaissance Christianity.[5]

According to an older version of the strong cosmological principle, the distribution of matter and energy at the heart of the universe results from pure chance and does not produce either place or preferred direction. Local irregularities, such as galaxies and the stars that constitute them, certainly reflect essential fluctuations of quantum origin that presided at the appearance of matter at the very first instants of the Big Bang. Little by little these fluctuations allowed the growth of irregularities in the temperature of the universe, such as those that were detected in 1992 by the American satellite COBE and that are, it is believed, at the origin of the large groupings of galactic families now observable. By nature, these original fluctuations were not subject to determinism, and the matter that today forms the universe could thus obey this statistical principle. Finally, the model allows that the universe is without boundary. In this case, there would be no separation between an inhabited part and an empty part, neither center nor any other privileged place. A well-known image illustrates this type of situation: the surface of a balloon, which can be traversed endlessly, with never an end to be found.

Within the framework of this model, there is thus no objective order in the overall universe, and its objective description, even a partial one, is impossible. For example, it would be illusory to want to describe objectively the configuration of the solar system in the universe, since there certainly exist, in the myriad of galaxies, many planetary systems that would answer to the same description, however detailed.

Briefly, Layzer's hypothesis denies the possibility of objective information or order in the universe at this scale of description. Every appearance of order, or of negentropy, at a lesser scale is thus a *creation.* Only the creation of negentropy at a lesser scale, if it exists, could be called objective. Here is an interesting thesis, which I discussed in *The Children of Time.*[6]

It is true that the validity of the strong cosmological principle in the preceding naïve formulation is less plausible today than it was several years ago. How can it be accommodated with the large walls of matter and the empty spaces that seem to occupy intergalactic space? Layzer suggests that to understand them, we must suppose not that the primordial universe was very hot, as the "standard" model proposes, but on the contrary that it was very cold (the appearance of a relatively high temperature in the universe 10 or 15 billion years ago, presently attested by the spectrum of the background radiation, could easily have resulted, after the Big Bang, from secondary processes such as the formation of the first large galactic clusters). But then what is the origin of the formidable pressure that must have destabilized this primordial universe; why did it begin to expand? On the other hand, the most recent measurements on the density of matter in the universe do indeed bring it closer, bit by bit, to the limit beyond which the universe could be closed, folded on itself like the surface of a balloon, but the measurements give a result that is still too low by about a factor of ten. It thus could be that the true geometry of the universe is very different from that of the model.[7] I nevertheless think that Layzer's thesis remains a precious guide, from the heuristic point of view, as to the original status of chance and information on the scale of the universe.

Let us now look to the no less audacious thesis of Roger Penrose. He remarks that all the major processes of the creation of negentropy in the universe are dominated by gravitation. "It seems that we have arrived at the conclusion that all the remarkable states of low entropy we can find around us . . . must be attributed to the fact that vast quantities of entropy can be won thanks to the gravitational contraction of the clouds of gas in stars."[8] Penrose explains that with a gas for which we overlook gravitational interactions, as Boltzmann did with his statistical theory of gases, the growth of entropy tends to render its distribution more uniform over the accessible phase space. For a system of bodies subjected to gravitation, however, the opposite holds. A high entropy is obtained by collecting those bodies under the effect of gravitation, and the highest entropy is attained by crashing those bodies into a black hole. The universe that surrounds us is a particular example of this. It began in a condensed state of very low entropy, linked to exceptional conditions summarized by the nullity of "Weyl's tensor" of general relativity, and is perhaps evolving toward a much more ordinary state of very high entropy, the Big Crunch. In such circumstances, we cannot hope to comprehend the second law of thermodynamics unless we amend classical physics to introduce true chance. For Penrose, that cannot be done except by constructing a quantum theory of gravitation endowed with the two essential ingredients of unpredictability and irreversibility. The current theories of gravitation include neither. Needless to say, the new theory Penrose wished to see created remains completely unformed.

TWO CONCEPTIONS OF CHANCE IN QUANTUM MECHANICS

Our two authors, anxious to propose a way in which to understand irreversibility and its link with cosmological expansion, agree on the idea that the key to this mystery is hidden at the heart of quantum chance. But these two authors have quite different conceptions of the role of quantum chance. Layzer insists on solidarity, a universal property manifested, par excellence, in Pauli's principle of exclusion. It is because of solidarity that he can propose his cosmological principle, that the universe is without microscopic information. For Layzer, the essential property of quantum mechanics is just that. It is impossible to attribute the least bit of objectivity to a ket that represents a quantum system if it is not the ket that potentially represents the whole of the identical quantum systems present in the universe. The interpretation of quantum mechanics in terms of ensembles does in fact allow certain obstacles and paradoxes to be solved, like those of Schrödinger's cat, but it does not resolve all the epistemological difficulties raised by this theory once an interpretation devoid of any reference to an observer is sought.[9]

In practice, Layzer proposes to attribute a cosmological origin to all quantum manifestations of natural chance. In effect, as we have just seen, the strong cosmological principle implies that a complete theoretical description of the universe cannot single out any specific physical system whatsoever. Thus, when we speak in quantum mechanics of the ket representing an unstable particle, for example, for Layzer we are actually speaking about a bit of a much more general ket that describes all unstable particles of the same nature, which are distributed in a more or less uniform fashion through the universe. Each member of the ensemble allows for the same representative ket, and the overall ket is the direct product of a practically infinite number of particular kets. Suppose that at the end of a certain time *t*, the probability that a particular particle has decayed is *p*. That means that a fraction *p* of the particular preceding kets in the direct infinite product must now describe decayed particles. But by virtue of the strong cosmological principle, we cannot know which ones have decayed. Thus the uncertainty as to the instant in which a particle decays is inseparable from the uncertainty as to its location in the universe. Quantum uncertainty about decay time and uncertainty about position in the framework of the strong cosmological principle are thus linked.

Obviously, in Layzer's view attributing any type of objectivity to the description of *one* quantum system by a ket or a wave function is unthinkable. For him, quantum mechanics can and does describe only collective, statistical ensembles of particles or identical systems, all being potentially innumerable in the immensity of the universe. This is a view that seems to me to approach Leibniz's conception of "monads."

All this has attractive aspects. On the other hand, it seems to me that the principal weakness of Layzer's approach is the fragility of the primary hypothesis. Does the strong cosmological principle have any chance of truly representing the universe in which we live? These past few years astronomers have gone from one surprise to another, so that we now envision substantially modifying the standard model that made the strong cosmological principle plausible. Firstly, there are great unexplained heterogeneities in the universe, such as those great walls and those vast empty spaces recently described.[10] Secondly, turning our instruments toward more and more distant galaxies and quasars, we see these objects as they were at an earlier and earlier time. Given the standard model of the Big Bang, which places the event about 15 billion years back, we expected that any objects we might find beyond more than 10 billion light-years would be younger and younger, witnesses to the birth of the universe. That expectation was in error. It is true that the inflationary scenarios allow us to have doubts about the true size of the universe. It could easily have been, over the fraction of a second that the inflationary period may have lasted, that the universe expanded over immense distances, much vaster than the temporal horizon of some 15 billion light-years: the light emitted by objects possibly situated beyond that horizon has not yet had the time to arrive at the Earth. Finally, there remains the fundamental difficulty of equating numbers that are so large that they are, for all practical purposes, considered infinite with numbers having the singular mathematical properties of infinity. Even within the framework of a strong cosmological principle and in a closed universe, without boundaries, the number of atoms and elementary particles that populate the universe remains thoroughly finite.

GOD: ANOTHER NAME FOR CHANCE?

Thus, for physicists, natural chance seems substantially different from the portrait painted by the mathematicians. There is indeed in nature a fundamental principle of independence and of equal a priori probability, as in the games of chance invented by humans. But the rupture between chance and nonchance is much more radical than those analogies might suggest. Physical chance imposes an absolute limit on reductionism beyond which all the models of that type can only fail. And yet that very rupture, far from leading to a chaotic world, engenders on the contrary a world in perpetual construction, in perpetual self-organization, from galaxies to bacteria. What a contradiction for mathematics!

Some propose calling this organizing principle "God." At first glance, what does it matter what it is called, as long as we know what is meant by the name? When we witness this return of God with the help of the science of chance, we cannot help but smile at how malicious history can be. Because God, far from contesting chance, as the Renaissance theologians would have had it, would instead use it to manifest himself. Or rather, God would seem to direct chance toward a destiny that only appears to be blind.

But to call this organizing principle by the name of God without descending into Spinoza's pantheism, we must endow it with a supplemental virtue, that of a final cause. Whether we like it, here is a breach that modern mysticism has not hesitated to rush to fill. The organizing virtue of physical chance does in fact evoke the idea of finality, as it was upheld by that irresponsible materialist the marquis of Lamarck or as it more recently inspired a more consistent mystic, Pierre Teilhard de Chardin, with his Omega point. In the fifties the words of this then famous Christian philosopher seemed to carry a whiff of sulfur and, if I dare say, inspired many progressive Christians by that very daring. The disdain in which he is often held today is caused by the modesty of his purely scientific contributions. But the audacity of his personalized and cosmological intuitions demands respect and perhaps still reserves for him the historical designation of precursor. But our vision of a natural chance cannot yet entirely wed itself to the Teilhardian vision of a universal force tending entirely toward a goal, that of calling matter to life and life to thought, since the trail of evolution shows evidence that the road has strayed too many times and accidents have been too numerous. Natural chance may perhaps be a constructor, but it is also, let us not forget, essentially opportunistic.

PART

4

FOR A PHILOSOPHY OF CHANCE

CAN WE UNMASK CHANCE?

14

Chance is like the emperor of a distant country in childhood tales. We talk about it, and it rules our lives, but no one has ever seen it. That may be what we used to take for granted, but is it actually true? Maybe there is a way to recognize it, to positively identify its presence when it does us the honor of dropping by. We have seen how difficult it is to measure probabilities, even for those who accept an objective interpretation of that notion. During this century, von Mises could not avoid detouring through the notion of "collectives" of events, that is, through the consideration of potential results of an infinite series of tests. Popper had to use all his genius to reduce that abstraction to the consideration of a finite series of trials, albeit of arbitrary lengths. With the works of Ray Solomonoff, Andrei Kolmogorov, and Gregory Chaitin, we now have a definition of chance that agrees with the intuitive idea that *in mathematics chance does not construct regularities or remarkable structures.* This is, however, still a negative definition. Something is said to happen by chance when it is not ordered or when it cannot be proven to be ordered. But let us take a closer look to see what this is all about.[1]

GIVE ME A RANDOM SERIES OF NUMBERS

We can imagine how Galileo might have decided to present this problem to his contemporaries. He would have called on Simplicio and Sagredo. Simplicio resolves to examine Sagredo's intelligence on this subject. He decides to prepare a test. In a notebook he writes, line after line, several equally long series of *T*'s (tails) and *H*'s (heads): "T T H H T H H H T T T H T H T T T T H T . . ." But from time to time, instead of inventing the series, he throws a coin and faithfully registers the heads or tails that result. When his page is full, Simplicio goes to see Sagredo. "Some of these series were the chance results of throwing a coin, and others were invented. Tell me, please, which are those due to true chance?"

Sagredo likes to rephrase practical problems in mathematical terms. These are more familiar to him. Living in the age of computers, he hurries to translate each series into binary numbers. The first series, for example, is written "0 0 1 1 0 1 1 1 0 0 0 1 0 1 0 0 0 0 1 0 . . . ," where each 0 represents a *T* and each 1 an *H*. In the present case this transposition is evident, although it could be a bit more complicated in other situations where there are more possible alternatives, but that does not alter the generality of the reasoning.

On reviewing the binary numbers line after line, Sagredo sees one like this: "0 1 0 1 0 1 0 1 0 1 0 1 0 1 0 1 0 1 0 1." Sagredo guesses that Simplicio has deliberately constructed this sequence by basing it on a simple principle: "Repeat as many times as necessary the sequence 0 1." If Simplicio showed him only one part of the series, Sagredo could easily guess how it continued. He would apply the well-verified adage that *natura non facit saltum,* which is the basis of all induction. In contrast, in the first series there is no obvious algorithm that can reconstruct the series or predict its continuation.

With this observation, Sagredo wants to tell Simplicio that the series he has just discovered must have been invented by Simplicio and not by chance. But does he have the right to say that? Indeed, he knows that a series of twenty throws can result in any one of 2^{20} possible series of 0 and of 1 and that each of those series has the same probability of "being thrown," whether or not it contains remarkable regularities. Thus, the two series cited as examples by rights have exactly the same probability of being realized, namely $1/2^{20}$.

But Sagredo thinks, doubtless with reason, that he should trust his intuition. What he lacks is a definition of chance, one confirming the idea that chance does not produce regularities of this sort (in mathematics, of course, since we have seen that in nature it is a different story . . .). The definition in question is that codified by Solomonoff, Kolmogorov, and Chaitin. Note that it is not *truly* a definition, except by default. It is a definition "by omission," since it is chance's opposite (order) that is operationally defined. Except in exceptional cases, this definition does not allow a decision as to whether a given series of numbers is due to chance or only seems to be. On the other hand, it lets us rule out the hypothesis that an ordered series such as Simplicio's "0 1" series is due to chance.

After thinking deeply, Sagredo proposes the following criterion: *the works of chance cannot be summarized.* To be more precise, he decides to attribute to chance only works in which *the information they contain cannot in any way be compressed.* Take a computer, for example. Every program this computer is capable of running is presented as a series of instructions—and in binary as a series of zeros and ones similar to Simplicio's series. There is no question here of the *meaning* or *purpose* of the program that is submitted. The computer blindly and mechanically executes its instructions, and that is enough. Many algorithms can fulfill a similar function (e.g., solve an equation of the third degree), but among these there is one or more that is shorter than all the others[2] and therefore cannot be compressed.

Let us return to Simplicio's series. In view of the criterion he had just adopted, the difference between the two examples cited becomes clear. The second series can be written "repeat the group '0 1' ten times. If we want to lengthen the series so that it contains not ten but a large number *n* of pairs of numerals, we would write "repeat the group '0 1' *n* times." The number of bits necessary to write such an algorithm is only a fraction of the number of bits of the true series, a fraction that becomes smaller as *n* becomes larger.

For the original sequence, however, there are no such shortcuts. The most economical way of describing the series is to write it out entirely. A computer charged with producing it would probably not be able to use an algorithm shorter than this: "print 0 0 1 1 0 1 1 1 0 0 0 1 0 1 0 0 0 0 1."

Let us adopt Sagredo's point of view. We will say that *a series of numbers is random if the smallest algorithm capable of producing it in a computer has about the same number of bits as the series itself.* Such is, in effect, the criterion proposed by Solomonoff, Kolmogorov, and Chaitin.

USE OF THE ALGORITHMIC CRITERION TO MODEL THE INDUCTIVE METHOD

The algorithmic definition of randomness does not replace the classical theory of probability. The evaluation of probabilities, the laws of large numbers, and the study of variations are still pertinent. But it does complete the theory, or rather extend it toward its foundations. As we will see, it allows linking the question of the existence of chance with that of undecidable propositions. The existence of such undecidable propositions is a result bequeathed to us by Gödel's incompleteness theorem.

The algorithmic definition of randomness also intersects the general problem of scientific induction, as shown by Solomonoff. In the model Solomonoff developed, a *theory* is no more than an algorithm capable of generating (reproducing) a series of observations that can always be transcribed in a series of numbers and of extending the sequence thus obtained by making predictions. Obviously, it is a slightly restrictive version of induction, but it agrees well with the approach of the empiricists and the positivists, such as Ernst Mach. This model also asserts that the shortest algorithm (the most compact theory) that reproduces all the observations should, a priori, enjoy the favor of theoreticians, in application of a commonsense principle of economy.

Within the framework of such a model, it is clear that inconsistent observations (obeying true chance) that consequently cannot be represented by an economical algorithm are not susceptible to being summarized by any theory whatsoever.

The notion of a minimal program is closely related to that of *complexity.* The complexity of a series of numbers is measured by the length (expressed in bits of information) of the minimal algorithm required to produce it. We can thus redefine chance by saying that *a series of numbers is random if its complexity is approximately equal to the length of the series written in binary notation.*

Given a group of numbers of n digits, it is theoretically possible to identify those with complexity of $n - 1$, $n - 10$, $n - 100$, and so on. The exact value of the complexity below which the series will not be considered random remains an arbitrary decision. It is to account for this restriction that I am obliged to introduce the word *approximately* in the previous definition.

Consider an example. The complexity of a series of n digits composed exclusively of zeros (or exclusively of ones) can be reduced to the length of a binary program such as "print n zeros," in which the element that takes the longest to write, when n is a relatively high number, is n itself. The complexity of the series of n zeros is thus approximately equal to $\log_2 n$. This is, indeed, the number of digits necessary to write n in binary notation. Of course, for a large enough n, $\log_2 n$ is very small relative to n.

Let us review natural whole numbers. In writing them in binary notation, we see that most of them have a complexity that is close to that of the numbers themselves. In other words, most whole numbers are represented in binary notation by a series that Sagredo would have declared produced by chance. Numbers whose binary coding is not compatible with the adopted criterion of randomness are exceptional. We shall see that the proportion of numbers of n digits of complexity p is just about proportional to 2^{p-n}.

We will keep only those numbers whose complexity is higher than a certain limit—for example, $n - 10$—as numbers representing a random series. We note:

—that there exist $2^1 = 2$ programs of length 1 capable of generating a series of identical digits: those that generate only 0 or only 1;

—that there exist $2^2 = 4$ programs of length 2: those that simply and unvaryingly repeat the combinations "0 0," "0 1," "1 0," and "1 1";

—that more generally there exist 2^k programs of length k, of which each generates by appropriate repetitions a different one of the series of numbers with n digits.

Thus the total of the numbers with a complexity less than $n - 10$ is $2^1 + 2^2 + 2^3 + \ldots + 2^{n-11} = 2^{n-10} - 2$. Therefore, there exist fewer than 2^{n-10} algorithms smaller than $n - 10$, each specifying a sequence of n digits. As there are 2^n different possible sequences of length n, the proportion of natural whole numbers shorter than or equal to *n* that fail to meet the criterion of randomness thus defined is about $2^{n-10} / 2^n$, or $1/1024$: *fewer than one thousandth of the natural whole numbers have a complexity less than* n – *10* and must thus be rejected by Sagredo as not having been produced randomly if he adopts the previous definition of randomness and applies it to the list of natural numbers written in binary notation.

I suggested at the beginning that to construct a series of numbers by chance, Simplicio could toss a coin and use the results. The coin toss is consecrated by tradition as the archetype of a physical realization of mathematical chance. Applying our conclusions in reverse, we can predict that if a coin is tossed *n* times to produce a unique series among the 2^n possible series (and thus any one of the natural numbers among the first 2^n numbers), there is a probability greater than .999 that the series of results actually represents a random series, but in one case in a thousand the series will appear extraordinarily ordered. In this sense, tossing a coin is *not* an infallible generator of random numbers.

THE LIMITS OF THE KNOWABLE

These properties of chance bring us closer to a general problem of epistemology: what is the limit of the knowable? Given a series of digits of length *n,* if we find a much smaller series of length *p* capable of reproducing the larger series, we can say that the number it represents is not a number constructed by chance. To prove positively that a given number was arrived at by chance, however, we would have to prove that there exists no algorithm shorter than *n* capable of reproducing it. Such a proof is unfortunately most often out of reach, as shown by Gödel's theorem.

The essence of this theorem can be formulated today by using the findings from the preceding paragraph.

It is once again Simplicio who will start us out. Today he proposes the following problem to Sagredo: "Find the smallest positive integer whose specification demands more characters than there are in this sentence." Sagredo sees that the sentence, which contains ninety-eight characters, should specify an entity whose specification requires more than ninety-eight characters. This is manifestly impossible. Let us generalize the problem in the familiar domain of algorithms written in the usual binary code. The sentence gives "find a series of zeros and ones whose complexity can be proven to be greater than that of the code we have just written." The program is supposed to calculate a number that no program of its size should be able to generate. The absurdity of the conclusion shows that in fact the program cannot work. Whatever might be its concrete form, it will never find the number in question.

Let us apply this result to the demonstration of theorems in a formal system of axioms. What is a formal system of axioms? It is precisely a series of irreducible fundamental statements in the form of minimal programs (if an axiom could be stated more compactly, that new way of expressing it would certainly be adopted as the basic axiom). Thus the information contained in a system of axioms satisfies the criterion of chance: it is expressed by a random series of digits. This system can be used to demonstrate the random character of other series, but only if the series studied are shorter than the formal system of axioms itself. Since all the minimal programs are represented by a random series, the theorem also implies that a formal system of complexity greater than n is needed to prove that a program aimed at constructing a series of n digits is minimal.

Insofar as complexity measures the random character of a series, the preceding result implies that in a formal system one cannot prove that a series was produced by chance unless the complexity of the number to be tested is smaller than that of the formal system itself. Of course, every formal system is necessarily of a finite length, while a series can be arbitrarily large. Thus there will always be numbers whose aleatory character cannot be proved. We encounter here one of the undecidable propositions to which Gödel drew our attention.

CHANCE AND CONTINUITY

The algorithmic definition of randomness obviously presupposes that the *concrete realizations of chance* have the power of natural numbers and not the power of the continuum.[3] In other words, the elementary events that underlie mathematical chance are enumerable, discrete events. The objection could be made that in the concrete world such a situation is exceptional and that most of the aleatory magnitudes of physics have the power of the continuum: position, velocity, potentials, electrical or magnetic fields, and so on. This objection has embarrassed more than one classical physicist, and mathematicians have taken great pains to extend the results of the theory of probability from discrete to continuous random variables. This is the subject of the theory of distributions, in which Laurent Schwartz notably excelled.

It is true, however, that Solomonoff's, Kolmogorov's, and Chaitin's theory restored relations with the old tradition of a congenital alliance between chance and a granular conception of the ensembles to which it applies, or as we say today, of quantification. In one sense the spirit of this theory is reminiscent of Ludwig Boltzmann's position, for Boltzmann ratified Pythagorus and Plato and recognized that whole numbers, and beyond them all things enumerable, have a particular reality truer than the continuum, which he viewed rather as a construct of the human mind. He was wary of the results obtained through differential analysis and did not fail to emphasize on several occasions that differential analysis is only a mathematical artifice allowing a "shortcut" relative to arithmetical methods.[4] Layzer also insists, with reason, on this fundamental property of the real: the presence of chance in the universe signifies a character who can play his part only on a granular field.[5] We know today that this view of things corresponds to a deeper reality: the fundamental theory that describes nature, quantum mechanics, is, precisely, fundamentally quantified.

The results obtained in the recent analyses of chance show us that in general it is impossible to prove that a number is random. In this sense, the actor chance of whom we speak is an actor in a theater of shadows. He is not seen directly, but the light of order and the stretches of emptiness it reveals unequivocally show him to us. Remembering Plato's famous cave, we find ourselves agreeing with him in order to state that if chance does rule the events of the real world, we here see only its shadow.

But our inability to prove that a number is random does not mean that truly aleatory numbers do not exist. In the same way, the fact that we cannot objectively prove that chance truly exists does not mean that it does not exist.

The impalpability of chance, even its evanescence, is a gift from the gods, since it leaves us free to pursue, everywhere possible, our ideal of determinism, of objective knowledge, and of domination of nature. Since we do not know where chance's realm begins, let us endlessly attempt to uncover order and its laws.

EINSTEIN, POPPER, LAKATOS, AND THE PROBLEM OF INDUCTION

Chaitin's precept clearly outlines the goal of the scientific enterprise: to find shorter ways of describing the universe. He does not tell us how the scientific researcher should or actually does proceed to do this. Analyzing the ways in which the scientist may succeed in wording such summaries is the subject, par excellence, of epistemology.

We have valuable testimony as to Einstein's thoughts on this subject. The letter he addressed on May 7, 1952, to his boyhood friend Maurice Solovine contains a striking summary of his ideas. Gerald Holton presents and comments on it in his book *L'Imagination scientifique* (The scientific imagination).[6] "Science," wrote Einstein, "is the attempt to coordinate the chaotic multiplicity of 'facts' into a system of thinking that enjoys unity and logical coherence." But how can we guess the system of axioms with which the facts should be confronted to see whether the former does indeed correspond to the sought-for summary of the latter? Induction, according to Einstein, does not follow a logical or necessary course. It is a free creation of the mind, guided only vaguely by facts. "There is no inductive method able to lead to the fundamental concepts of physics," he observed elsewhere. The misunderstanding of this state of affairs was the fundamental philosophical error of many researchers of the nineteenth century.[7] Once proposed, the axioms allow us to deduce their logical consequences, which must be compared with experimental findings. The latter will be the results of precise experiments, designed to test the former. This practice itself partially escapes logic, since, as Einstein recognized, "The relationships between concepts intervening in [the axioms] and the experiments are not of a logical nature." Still, if the theoretical predictions are corroborated by the experiments, our confidence in the rightness of our construction will be reaffirmed. We are in no way dealing with proofs, for the three following reasons: firstly, correct predictions can be drawn from fallacious axioms; secondly, to prove a theory once and for all, it would have to be submitted at one time to an infinite number of observations; and finally, the experimenters charged with verifying one or the other aspect of a theory are themselves fallible, however honest and scrupulous they may be. We must therefore be content with a presumption of veracity, attested by the two following criteria: firstly, the theory must not contradict the experimental findings; secondly, the theory must have the aesthetic qualities of simplicity and harmony. We can compare this second criterion with the idea from Chaitin's theory that a theory is the more powerful the more "ink" it saves us in describing the world.

The idea that a theory can never be proved but only sometimes purposely falsified lies at the heart of Karl Popper's epistemological thought. This somewhat negative idea in no way kept Popper from believing that "science must strive to discover the objective truth, a truth that depends only upon facts, which is above the arbitration of human beings, and certainly above intellectual fads."[8] Affirmation of this faith is not rooted in any demonstration. The works of contemporary epistemologists never cease to expand the catalog of difficulties that arise in pursuing this program.

Among these difficulties, one of the most important, already suspected but not discussed by Einstein, is the fact that there is nothing like a raw observation or a raw experimental finding that is not already partially a product of an underlying theory.[9] These same contemporary thinkers also admit that the stated ideal—that a theory is more powerful the more it condenses facts—although theoretically true, does not correspond in practice to what scientists consider to be a particularly fruitful hypothesis. "It is wrong to believe that the falsification of bold conjectures, which are highly falsifiable, represents a significant advance in science," wrote Chalmers. "Some theories take the form of bold, risky conjectures, while others are prudent conjectures whose implications do not seem to contain significant risks. If a prudent conjecture fails testing by observation or in an experiment, it would be proven false, whereas if a risky conjecture passes such a test, it would be said to be confirmed. The significant progress takes place with the *confirmation* of bold conjectures or the *falsification* of prudent conjectures."[10]

The conduct of scientific researchers is analyzed in even more detail by Imre Lakatos,[11] who emphasizes how scientists, who do not want to risk a hypothesis in vain, generally surround the systems of axioms then in favor (because of their past successes) with a shield of *protective* auxiliary hypotheses that allow them, in cases of a conflicting observation or experimental result, to save the essence of what they believe has been acquired and thus to avoid, except in very rare circumstances, radical requestioning or revolutions in scientific thinking.

A sociological observation of the scientific community certainly confirms these reservations overall. If it emphasizes the concrete difficulties of the exercise of science, such an observation does not detract from the merits of Einstein's reflections on the quintessence of the art of elaborating a scientific theory, any more than it contradicts the bold speculations of Solomonoff, Kolmogorov, and Chaitin. These latter uncovered with rigor the general forms that distinguish the scientific enterprise from other human enterprises, sculpted in the unformed clay of the "chaotic multiplicity of experience," to use Einstein's own words.

THE QUESTION OF FREEDOM

15

After the question of existence, that of freedom—of my personal liberty as a human—is unquestionably the central question of philosophy. I truly wish to know to what degree I "myself" am able to decide my own actions and to determine my own behavior in any area whatsoever. Or even whether this claim to free will is only another illusion whose conceit I must finally admit . . .

Asked in this way, the question is much too vast to be adequately addressed in this book, even with the help—which I will not hesitate to accept—of some of the great philosophers. I will more modestly try to clarify it from the perspective of the existence of an essential chance. If the real itself is awash in contingency and riddled with unexpected events, doesn't this open the door to the possibility of an "I" who can make decisions? There are, however, still vast distances between accepting true chance and verifying the fundamental metaphysical hypothesis of a free self. Contemporary science, while allowing us to specify the terms of the debate, does not truly enable us to bridge this gap.

Certainly the question of the existence of an intrinsic chance, which I am defending here, and that of liberty are not on the same level and must be philosophically distinguished. In the last century Maxwell, Peirce, and James were ardent apostles of indeterminism as a necessity in preserving human liberty, but their arguments can seem fairly vague today. Maxwell seemed to confuse liberty and sensitivity to initial conditions. Peirce categorically rejected, in the name of *becoming*, a causality whose ability to explain nature is undeniable. James searched the realm of psychology for explanations that were then visibly beyond the reach of that discipline. This does not mean that chance and liberty may not be linked in some way, since it is difficult to imagine the existence of free will without the existence of some type of chance.

Curiously, the philosopher who wanted to base human free will on true chance would be as poorly armed as the determinist when attempting to answer the crucial question of personal responsibility. In forging too close a tie between the existence of free will and chance, what place would there be for responsibility, which is the only truly dignifying notion?

THE TEMPTATION OF DETERMINISTIC MATERIALISM

Since Montaigne, the fortunes of liberty and chance have been linked. As Martin Daniel wrote in summarizing Montaigne's thoughts, "It is precisely because there is a certain distance (chance) between God and man that man can assert his freedom." And he had this nice formula: "The function of chance is to allow us to bargain with God."[1] The emergence of classical determinism, denying the reality of chance, would naturally relegate the problem of free will to that of an impossible philosophical problem. Spinoza saw no other solution to this dilemma than to propose the following idea: liberty is nothing more than the name we give to our internal determinations; the feeling of being free consists in acquiescing to these determinations. Despite Kant's critical analysis restricting the field of scientific determinism to the sphere of phenomena and opposing them to true reality, the tendency until the middle of the nineteenth century was to oppose determinism to the experience of free will. The discovery of the general conservation of energy at the middle of the nineteenth century allowed materialists to whet their knives against spiritualists and other defenders of free determination. The regularities unveiled by the theory of probability gave them additional ammunition. In his *History of Civilisation in England,* published in 1857, Henry Buckle used the regularities observed in criminal statistics against the idea of free will. Since, in English society, the number of crimes in both good and bad years remains near an average value, shouldn't we conclude that these acts are in some way inevitable, that the criminals cannot escape that subtle determination we call the laws of chance? Physicists, with Maxwell, emphasized that probability theory is bound to apply to the social sciences in the same way that it applies to gas molecules. The extension of "scientific" materialism then caused materialistic determinism to reach a sort of apogee that lasted well into the twentieth century, producing characteristic works such as Lenin's *Materialism and Empiriocriticism.*

Even as early as the middle of the nineteenth century, however, a reaction began to appear. Maxwell himself remarked that we cannot know the causes of molecular motion in all their detail. He saw in this practical limit a possible reconciliation with free will. The type of indeterminism he proposed at the level of the most minute of molecular mechanisms might, according to him, enter into play each time a system reaches a point of instability, as in the case of a stopped top that cannot "know" in advance onto which side it will fall.

Presented in this form, the argument seems to lead more toward the development of a theory of unstable complex systems sensitive to initial conditions, to which Poincaré devoted a great deal of thought. Those reflections led not to the restoration of an essential chance but to the theory of deterministic chaos, which can be of no help in our problem, except to insist on the illusory character of our feeling of liberty.

For the dedicated determinist, the fundamental difficulty does not necessarily lie in accommodating determinism with liberty. Spinoza's old argument remains entirely valid. It is enough to redefine liberty in sufficiently constrictive terms, as that author so well knew how to do. If liberty is no more than the art of acquiescing to our internal determinations, there is no conflict between determinism and the feeling of liberty. But who can be truly satisfied with such a solution? It cannot please us, since in truth humans would then be neither truly free nor truly responsible.

These are not the only objections to the temptation provided by materialistic determinism. It is interesting here to remember those counterarguments formulated by Karl Popper, an epistemologist and theoretician of science who did a lot to clarify the problems raised by the notion of chance and the calculation of probabilities. Popper, tempted in his earliest writings by a deterministic and purely materialistic approach, was not able to resolve within the framework of that philosophy the cosmological problem "of understanding the world, ourselves, and our knowledge, considered as a part of the world." Popper's argument was founded on a sort of infinite regression of our knowledge of reality based on temporality. Since we are a part of the world, we cannot predict the discoveries still to come (or they would already be contained in our description of the world). The description, and the prediction, of our scientific knowledge can never be total, since it cannot include knowledge that we are *just then in the process of acquiring.* Thus we cannot totally predict the future state of the world.

In this way this person who devoted so much effort to discerning the existence and properties of true chance ultimately could not avoid proposing thought's autonomy from its biological substrate, the brain. His philosophical system finally split reality into three communicating but distinct worlds: the world of physical objects, that of feelings, and finally that of organized thoughts and scientific theories. The reason for this recourse to different worlds can be found in the idea that physical indeterminism is not enough to justify free will. All that it allows is the creation of the space necessary for free and creative human actions.[2]

CHANCE, ESSENCE, AND EXISTENCE

The existentialists tended to merge the two questions of chance and liberty into a single question and to relegate liberty to a primary condition of existence. Not satisfied to be a simple acquiescence to our internal determinations, as suggested by Spinoza, our freedom can find expression by leaning on chance. Here, in a nutshell, is the thesis advocated by Jean-Paul Sartre.

Sartre's existentialist philosophy takes the position that "existence precedes essence." To exist, therefore, means to exist in a state of contingency, at liberty, in the uncertainty of becoming. The world does not have a given meaning. Beings, existing in potential liberty, must give it that meaning. This explains the importance Sartre gave to the distinction and analysis of the relationships between the *en-soi* (in-itself) and the *pour-soi* (for-itself). Freedom is to be found not in the *en-soi* but in the *pour-soi.* The two notions must in fact be carefully distinguished.

> Freedom comes from that fact that human reality lacks essence because essence is gained through a perpetual battle against itself, and what it was is separated by a great chasm from what it is and from what it will be. It is free, finally, because its present being itself is meaninglessness in the form of "reflection-reflecting." Man is free because he is not self but attendance at self. He who is what he is cannot be free. Liberty is, precisely, that nothingness which *is* and *was* at the heart of man and which constrains human reality to *become,* rather than *to be.*[3]

In this text Sartre addresses the question of the status of being itself. He makes something essential of the association between being and liberty, and in liberty it is nondeterminism that is essential. Briefly, he refuses to define being other than by way of liberty. Being is not being entirely.

It seems to me that *not being entirely* is impossible except in a world inhabited by an essential chance and that from this point of view, modern science does not disagree with that approach. However, espousing this philosophy is perhaps not enough to answer the very questions it raises. What is the nature of that liberty, the foundation of existence? What are its relationships with materiality? Sartre saw liberty and existence as issuing together from the same spring. In the framework of a physical theory stripped of a too-rigid deterministic yoke, the idea of indivisibility between liberty and existence awoke new echoes. Thus, questioned by Guitta Pessis-Pasternak, who asked him where free will was located, Henri Atlan answered:

> Probably in the interaction between genetic determinisms and the determinism of the environment, insofar as it is through these interactions themselves that individuality is constituted. Freud's formula: "Where the 'that' was, the 'I' must arrive," although expressed before the discoveries of molecular biology, reflects a reality that is just as valid since. 'Free will' does not so much concern the range of maneuvering as it does the constitution itself of the subject, the fact of constituting itself as an individual.[4]

This thought, with its Sartrian accents, was expanded by the psychoanalyst Miguel Benasayag:

> Liberty was for a long time identified with the notion of free will: man was free to choose the right path (of course preexisting and to be discovered) or of fooling himself and thereby condemning himself . . . This perspective, unaware of the symbolism and the role of the "event,"[5] remains necessarily limited to the field of ethics, of "facts.". . . But if we renounce determinism, we can find another meaning [for the concept of liberty]. "Relative progress" can be the development of a *construction project,* in the framework of a system that is relative and contingent.[6]

Contrary to Sartre and to the neoexistentialist philosophers, stirring up these problems in the search for an adapted response, philosophers such as Kierkegaard and Bergson saw no other way out than to invoke a transcendence: something whose existence escapes ordinary materialism.

LIBERTY AND TRANSCENDENCE

If Søren Kierkegaard is justly considered the founder of existentialism, it is because he too proposed liberty as the foundation of being. For him, however, liberty is the mark of a (transcendent) supernatural quality in human existence. In some ways we find here an extreme form (insofar as it surpasses materialism) of the Sartrian theme of liberty as the basis of existence. We must recognize that this philosophy brought Kierkegaard neither ease of spirit nor serenity. The reason for this is the very transcendence that is evoked, so lofty that it is forever inaccessible. One cannot reconcile oneself with one's transcendent self. Transcendent reality being outside time and time being inseparable from the world, the only point of contact between the two realities is in the present instant. In fact, Kierkegaard considered the present as the point of contact between the infinite straight line of transcendence and the tangent circle of the world. But how fleeting and ephemeral is that fusion!

Another philosopher's view of the problem of freedom is that of Henri Bergson. Like Kierkegaard, Bergson proposed a dualistic solution to this problem. But he argued for the existence of a much larger zone of contact between the material, rational world and the intuitive, spiritual world. For him, the same organ apprehends the two faces of reality, through the two different functions of intelligence and intuition. Intelligence, for Bergson, is the faculty of reasoning. It ensures a parallelism between material reality, with its necessity, its determinism, and the brain. Intuition, on the contrary, represents in a way the other side of the intellectual faculties, the mind, which allows apprehension of transcendence, human liberty, and other faculties of the brain that the physicobiological scientists will never be able to explain. Physical laws can be apprehended by intelligence, but surprises, the unforeseen, the frescoes of *becoming* that are being assembled—these are accessible only to the intuition. It is from that side and through that instrument that we can hope to apprehend the reality of disorder.[7]

Bergson believed in a continuous creation. He saw in the actions of organization in biological evolution "something explosive" founded on a fundamental chance. But the unpredictability of nature did not lead him to despair. On the contrary, it brought him the joy of knowing himself to be free. These are the magnificent pages taken from the *Essai sur les données immédiates de la conscience* (1889), of which I must be excused for quoting a rather large extract, since the words resonate like a hymn to joy:

> Free acts are rare, even on the part of those who are most used to observing themselves and of reasoning about what they do. . . . When our staunchest friends are in agreement in advising us to do something important, the feelings they express with such insistence come to rest upon the surface of our *self* and solidify there in the manner of the ideas of which we were just speaking. Little by little, they will form a thick crust that will blanket our personal feelings; we will think we are acting freely, and it is only when looking back on it later that we will recognize our error. But also, at the moment in which the act is going to be accomplished, it is not rare that there would be a revolt. It is the suppressed self that surfaces. It is the external crust that breaks, ceding to an irresistible push. There were operating thus in the depths of this self, and below those very reasonable arguments that were juxtaposed, an agitation, and even a growing tension of feelings and of ideas, probably not at all unconscious, but to which we did not want to pay attention. In thinking deeply about it, by examining our memories carefully, we will see that we ourselves formed these ideas, we ourselves lived these feelings, but that, through an inexplicable reluctance to will, we had pushed them into the depths of our being each time they emerged at the surface. And this is why we seek in vain to explain our abrupt change of mind through the apparent circumstances that preceded it. We want to know by virtue of what reason we decided, and we find that we decided without reason, even perhaps against all reason. But it is precisely this, in some cases, that is the best of reasons. Because the action accomplished then no longer expresses some superficial idea that could almost be external to us, distinct and easy to express: it responds to the totality of our feelings, of our most intimate thoughts and aspirations, to that particular conception of life which is the equivalent of all our past experience, briefly, to our personal sense of happiness and of honor. Also we were wrong, in order to prove that man is capable of choosing without a motive, to go seeking examples in the ordinary and even indifferent circumstances of life. We would show easily that these insignificant actions are linked to some determining motive. It is in solemn circumstances, when it is a question of the opinion we will give of ourselves to others and above all to ourselves, that we choose, perhaps even against what we might call a motive; and this absence of

> any tangible reason is so much more striking as we are more deeply free. . . . Briefly, we are free when our acts emanate from our entire personality, when they express it, when they have, with it, that indefinable likeness sometimes found between the work and the artist. In vain, one would allege that we are then surrendering to the all powerful influence of our character. Our character is still us; and because we have liked to divide the person into two parts, in order to consider successively through an effort of abstraction the *I* that feels or thinks and the *I* that acts, there would be some puerility in concluding that one of the two *I*'s applies pressure to the other. . . . In a word, if we agree to call any act that emanates from the *I* and from the *I* only free, the act that carries the mark of our person is truly free, since only our *I* can claim paternity.[8]

Here Bergson points to the heart of the difficulty that plagues us: the freedom in question cannot be confused with the stumblings of the drunken sailor or the lack of responsibility of a computer programmed to respond by chance to any question asked. It has to do with the essential acts of our lives, those that engage our personalities and shape them at the same time. In this way it is inseparable from the question of the materiality or immateriality of thought.

We know the nature of Bergson's response to this question. Nonetheless, although Bergson was up to date with the latest findings of the psychological and cognitive sciences of his time, those findings have obviously changed considerably since then. Hasn't recent philosophy tended to transcend the closed fields of professional philosophers and become a preferred activity of scientists? This movement was already becoming apparent in Bergson's time, with the passionate debates surrounding the birth of quantum mechanics. Schrödinger's last works clearly proclaim it: modern philosophy, having tempted the physicists, now attracts biologists. It remains to us to weigh the meager crop both have harvested so far.

A FALSE TRAIL: QUANTUM INDETERMINISM

The apprentice philosopher-physicists who witnessed the development of quantum mechanics, with its fundamental indeterminism, thought they had found in it a solution to the problem of freedom. It was, of course, a false solution. First, it confused two levels, that of human consciousness and that of the atoms and electrons. Can one seriously evoke an electron's free will? This would be at least as risky as attempting to explain the formation of the galaxies using an understanding of the whirlpool caused by a draining sink. In addition, and even more dangerously, that solution mistook the nature of the uncertainty introduced by quantum mechanics. Neither liberty nor indeterminism exists for an electron; there is only an indetermination for us on the subject of the electron's true nature. The electron is not free to behave as a wave or a particle: its nature is to be neither what we call a wave nor what we call a particle. The "liberty" of the electron, insofar as it is related to the question of quantum indeterminism, is not an ontological freedom but an indetermination of our discourse or a weakness of our categories in a given experimental context. Certainly it is not the type of liberty the philosophers would wish to claim for humans.

Sixty years after the passionate debates of the "Copenhagen school" and the proposal by some philosophers to base liberty on the new properties of electrons, Bernard d'Espagnat, in measured terms, makes the case for the meaning of contemporary physics in the problem of liberty. In his book *Une incertaine réalité* (*Reality and the Physicist*), he states that quantum mechanics teaches us that science deals not with the real in itself but rather with phenomena—that is, the real in its relationship with consciousness (thereby confirming Kant's intuition). In this closed field, scientific practice seeks to enclose phenomena within laws. What is more natural than that there would be no place in this approach for liberty? But modern science also teaches us its own limits. Its field does not cover the real in its totality.

> The so important role that pronouncements of only *weak* objectivity [i.e., those requiring the presence of an observer, even though they are valid whatever the observer] play in basic physics of our times and the difficulty (which we have good reason to think might be insurmountable) there is in replacing them unambiguously with objective statements in the strong sense are serious indications in favor of the conception that science—and in truth not only elaborated science but also that elementary science that is common sense—itself forges its own concepts directed above all at being of use in description.[9]

We are thus not forced automatically to adopt a dualistic solution relative to this problem of liberty. There is no doubt still room for more nuanced positions, because in the end, we do not know what reality is in itself.

THE UNCERTAINTIES OF CONTEMPORARY BIOLOGY

This moderation seems to me to be confirmed by an examination of contemporary findings in biology, and particularly in neurobiology and other sciences that concern the brain. Taking stock of all the philosophical attitudes, all the "isms," in their confrontation with current neurobiological knowledge, Paul M. Churchland cannot discredit even one of them—not even the extreme position of John Eccles, the 1963 Nobel Prize winner in physiology and medicine and a convinced dualist, although Churchland visibly leans toward a unirealist interpretation, identifying thought with physicochemical cerebral activity.[10] Without going into the details of the numerous possible variants, we can follow him in distinguishing four overall attitudes according to the response given to the following question: *Do you think that the concepts of ordinary psychology (joy, sorrow, pain, etc.) could be one day explained by experimental neuroscience and reduced to or identified with physicochemical mechanisms implicating the activity of neuronal groups?* In response to that question, some individuals predict that psychological concepts will one day be reduced to and explained in terms of neuronal mechanisms. These I will call *reductionists.* Others maintain that such a reduction is unthinkable, since human behavior has a spiritual source, or at least a nonphysical source. These are *dualists. Functionalists* say they are interested only in the visible effects of thought, not in its nature. They agree with the dualists in judging the reduction of mental states to neuronal states to be impossible, because they understand the enormously diverse neuronal states potentially leading to one and the same observable behavior. The fourth group, the *eliminative materialists,* judges that current psychology is still much too empirical, imprecise, and approximate for the concepts it forges to lead to valid scientific analyses.

We have seen that Eccles leaned toward a clean break between thought and its physiological underpinnings, despite the obvious relationships that thought maintains with the brain. His philosophy is thus overtly dualist.[11] Another Nobel Prize winner, Gerald Edelman, leans toward a realism strongly tempered by the observation that both the brain and thought are the products of a long selective evolution. This observation makes possible, in his view, a sort of *emergencism* that is less distant from the dualist position of Bergson than from reductionist materialism.[12] Also an epistemologist and researcher in the neurosciences, Patricia S. Churchland reviews the most recent knowledge and comments on interrogations into the nature of consciousness in her remarkable work *Neurophilosophy.* She observes that consciousness must be distinguished from perception (many clinical reports cite obvious examples of unconscious perception, unknown to the subject) and, in reference to the problem of the reduction of mental states to neuronal states, shows that over the past few years, we have been witnessing a *coevolution* of neurobiology and psychology, letting us hope that they will converge, not in the immediate future, but in a more or less distant future.[13] In France Jean-Pierre Changeux does not hesitate to defend, in *L'Homme neuronal* (*The Neuronal Man*) and *Matière à penser* (*Conversations on Mind, Matter, and Mathematics*), a vision of thought that is frankly reductionist without letting himself become trapped by the difficulty raised by such a view: how can a brain "which would produce thought like the liver produces bile" (as Georges Cabanis said in 1801) engender attitudes that are sometimes contrary to the interest of the individual, and even to the survival of the species? The truth is that, although we know how to decipher certain elementary mechanisms of perception, such as the response of a cortical zone to specific stimulation, we have not the faintest idea as to what conscious representation is or—supposing that such representation exists—in which part of the brain it takes place. Thus, the American neurophysiologist Benjamin Libet was able to show that the individual corrects the sensations received in order to "reframe" them temporally, as a function of the time of propagation of the nervous influx from the sensory organ concerned to the "brain." A lively debate surrounding these works and their interpretation has once again shown us how total is our ignorance of the spatial and temporal framework within which these recalibrations take place.[14]

■

Thus, at the end of this slightly eclectic review, we must admit that, in the realm of neurobiology, we have above all been made aware of the extent of our ignorance. We are still a long way from understanding the mechanisms of thought well enough to be able to resolve, or even usefully clarify, the question of human liberty.

Our quest is indeed forced to stop on this slightly disillusioned note. Yet the chance that inhabits nature is one of the preconditions making free will possible. No, free will cannot be understood simply by pointing to that natural property. Yes, neuroscience will one day allow us to discern this debate more clearly, by making available constructive elements. No, it cannot yet do so. Thus, today free will remains above all a metaphysical question, just as much as the problem that is linked to it: that of its relationships with personal responsibility. Both may still require the attentions of psychologists and metaphysicians for a long time to come.

ON THE JOY OF NOT KNOWING

16

If everything could be known, would life be worth living? If a fairy godmother offered us, on our eighteenth birthday, the book in which the whole of our future lives were written out in excruciating detail, would we call her a good or an evil genie? Hope can be so much more beautiful than certainty . . .

It is impossible to deny the essential role that the prod of uncertainty plays in existence. That role is obvious, for example, in affairs of the heart. Nothing is more delicious than the first moments in which one feels that one has set out on a new road, where new relationships are formed with someone who is still an unknown and whose reactions appear unpredictable and disconcerting. What a marvel to feel oneself on the frontier of an unknown country. The tender complicity established when uncertainty about the other's love is resolved does not protect this mixture of love and fear from further surprises, day after day. Literature is full of romantic illustrations of these truths we all live.

From this comes the joy of not knowing, which is even more beautiful when it is shaped by real uncertainty, caused by the existence of a true chance, and not only by a destiny that is already written but hidden from us.

And yet, as the history of human thought teaches us, the yawning chasm of uncertainty does not necessarily lead to joy. It can also bring despair. Neither of these two routes is truly indicated for us by science, but both are a part of humanity's inheritance. For this reason, I will bring them up before finishing this work.

DETERMINISM AND DESPAIR

René Thom, the modern bard of determinism, does not understand how the supporters of a new approach to science can reserve a place for chance in the universe. In the preface he wrote for the reprinting of Laplace's *Essai philosophique sur les probabilités,* he superbly satirized the hope—laughable, in his view—of preserving liberty in this way.

> The modern disparagers of determinism, disciples of chaos and of the *Scienza Nuova,* theoreticians of order through noise, of order through fluctuation (I. Prigogine), have sung the death of determinism and exalted the aleatory. This in the desire to safeguard the possibility of radical innovation (in particular, to allow for human freedom). . . . In science, the purely aleatory is the Markov process, where all trace of the past is abolished in the genesis of the new; at each step the creative "big bang" is reiterated: the purely aleatory demands a fact without cause, that is to say, an *absolute beginning.* . . . If one wants to make of the aleatory, of the stochastic, an object of science, it is necessary to submit that arbitrary creator to constraints, to laws. . . . Quantitative science constrains its creator God to the repetition of a gesture—always the same one—aimed in any direction whatsoever. It reduces God to the Brownian stagger of the drunken sailor. What a pitiful attempt to dodge determinism is this vision of an insane God eternally constrained to the burlesque alternatives of Buridan's donkey. Is this any way to preserve the possibility of radical innovation, of creative invention?[1]

Thom's mistake, to my mind, lies in his not having perceived the difference, minimal in appearance but colossal in its consequences, between physical chance and mathematical chance. Yes, the existence of chance presupposes the possibility of an endless repetition of a single gesture, here and everywhere. But no, this repetition does not begin anew at each instant through the limited and idiotic choice of Buridan's donkey. In contrast to mathematical chance, physical chance constructs and deploys itself. Beyond being merely the absence of memory in the Markov process that Thom invokes, chance accumulates a formidable ability to repudiate this sense of the absurd in the storehouses of matter and life.

CHANCE AND ABSURDITY

Certainly no one more than Albert Camus has appreciated the absurdity of the human condition insofar as it is the plaything of a chance devoid of creativity or meaning. I cannot resist quoting here from some of the pages in which this absurdity appears in its earliest incarnation, even before it is tempered, in Camus's later writings, by his political and personal social involvement.

As early as 1942 Camus expressed himself thus in *Le Mythe de Sisyphe* (The myth of Sisyphus): "I said that the world is absurd, and I went too far. The world in itself is not reasonable, that is all that one can say. But what is absurd is the confrontation of this irrationality with that desperate desire for clarity whose call echoes in man's deepest depths. Man is as responsible for the absurd as is the universe."[2]

Later, in *L'Homme révolté* (The rebel [1951]), he rises in rebellion. "For twenty centuries, the sum total of evil has not diminished in the world. No second coming, either divine or revolutionary, has occurred. Injustice still attends all suffering, even that most deserved in the eyes of men."[3]

The solution Camus proposed, in the name of human dignity, is well known: to revolt. Camus was, of course, neither the first nor the only philosopher to have thrown himself body and soul into the problem of the absurd, and his completely personal answer, delving into the forces of life itself and into human rebellion—the only possible answer, in his eyes—is after all only one among many possible answers. Rebellion and then commitment among one's own was the only answer that seemed to allow him to reject the solution of suicide, which preoccupied him on numerous occasions. This revolt is based on the evidence of the absurdity of life, abruptly brought about by the shocking revelation of death: "Human insurrection, in its highest and most tragic form, is and can be no other than a long protest against death, a furious indictment of that condition led by the generalized sentence of death. . . . The rebel does not demand life, but reasons to live. He refuses the consequences that death brings. If nothing lasts, then nothing is justified, because whatever dies is meaningless." Jeanson's biting criticism during the Camus-Sartre dispute is as mean as it is unjustified: "God concerns you infinitely more than do men," which meant that Camus was no more than a *petit bourgeois*, more preoccupied with abstract metaphysical questions than with the defense of the downtrodden of whom he spoke so much. But Camus explained himself:

> In that way rebellion proves that it is the very movement of life and cannot be denied without renouncing life. Its purest cry, each time, brings a being to life. It is thus love and fertility, or it is nothing. Revolution without honor, the revolution of calculation that, preferring an abstract man to a man of flesh, denies the living individual as often as necessary, puts resentment in place of love. Once the revolt, forgetting its generous origins, lets itself be contaminated by resentment, it denies life, it courts destruction and raises the giggling cohort of those small rebels, future slaves, who end up being offered for sale, today, in all the markets of Europe, into whatever type of servitude. . . . We understand then that revolt cannot survive without a strange type of love. Those who find no respite either in God or in history condemn themselves to living for those who, like them, cannot live: for the downtrodden. The purest moment of rebellion then will always be crowned with the agonized cry of Karamazov: "If they are not all saved, then what good is the salvation of one alone!"[4]

THE PHILOSOPHERS' REBELLION

To Sartre, who accused him via Jeanson, Camus replied: "Unlike you, I have not turned my armchair to the flow of history"—a clear allusion to Sartre's period of companionship with the historical materialists and communism. But this is an element that is extraneous to existentialism.

Among the existentialists, and with Sartre in particular, the answer to the problem of the absurd is anything but diametrically opposed to that of Camus. For Sartre, existence precedes essence, which means that he affirms the primacy of liberty over the construction of the self and the world. On the whole, Sartre's existentialism is perhaps less pure than that of Camus, insofar as it is hybridized with historical materialism, or more precisely, with the idea that history has a direction. The two philosophers agree again, however, in the conviction that the universe has no meaning sui generis and that it is we humans who must find one. Sartre's classic distinction between the *en-soi* (in-itself) and the *pour-soi* (for-itself) reflects this idea. The *en-soi* is the material world, which is endowed with existence but has no essence. As we have seen, the *pour-soi* is the being that has existence but must take itself in hand to give meaning to its existence. This takes shape in the famous metaphor of the waiter in the café who exists only insofar as he plays his role of waiter in a café. We all play parts, in accordance with the idea that we make ourselves what we have to be. And it is this role we play that ends by transforming the *en-soi* into the *pour-soi.*

With Heidegger we see appear, in the direct line of existentialism, a circular definition of existence in the form of a word that is at the same time a loop onto itself:

Being-in-the-world
↑ ____________ ↑

Being exists only in the world, and the world exists only through Being. Unfortunately, Heidegger seems to have delighted in obscure speech, and his largely coded language can be understood only by initiated philosophers. It is nevertheless with Heidegger that the analysis of complexity appears succinctly, inseparable from those retroactive loops that characterize it. For Heidegger, existing means, as the word's etymology suggests, being-beside, with a loop like that indicated in the diagram. In existence the idea of Being and the idea of the world are inseparable. One cannot be if one is not in the world. It is thus in our *relationships* with the world, the external phenomena, that existence itself is constructed.

Heidegger is today highly criticized for his public life, especially his political attitudes. Just as the commitments of Camus and Sartre are judged by most of our contemporaries as being quite respectable, so those of Heidegger are judged ambiguous because of his flirtation with Hitler's national socialism. Realizing that great thinkers, such as Heidegger, have been able to compromise their philosophies of existence with their everyday lives and be so wrong can only be a cautionary lesson about the mind's ability to guide us through life and should cause us to take great care with regard to all the "great thinkers" of yesterday and of tomorrow.

TRANSCENDENCE AND DESPAIR

Kierkegaard, the philosopher of liberty, was determined to find a single answer to the absurdity of life. He thought he had found it in transcendence. Having voluntarily sacrificed his loves to research into the meaning of life,[5] in 1849 the melancholy philosopher published a essay on despair, *The Sickness unto Death.* There one can read a new and personal analysis of the relationships between anguish and time. As M. Grimault commented: "Faced with the past, a man can feel regret, remorse, repentance, but not anxiety. Anxiety only appears when faced with an indeterminable possibility: the future."[6]

In this treatise, he defines chance and necessity as two sides of the same coin in the construction of the self: "For the purposes of becoming (and the self must become itself freely) possibility and necessity are equally essential. Just as infinitude and finitude (*apeiron peras*) belong to the self, so also do possibility and necessity. A self that has no possibility is in despair, and likewise a self that has no necessity."[7]

But why this search toward transcendence? In *The Point of View for My Work As an Author,* Kierkegaard himself explains the path that led him to it: "I was so profoundly shaken that I understood the absolute impossibility that I, as I was, could find the peaceful and assured milieu where most men lead their lives. I needed either to throw myself into despair and sensuality or to definitively choose religion as the only thing necessary: the world to an extent that was cruel, or the cloister. Basically, it was decided that I would opt for the latter, I could do no other; the eccentricity of the former simply expressed the intensity of the latter."[8]

Here I must emphasize the necessary link Kierkegaard saw between despair and our condition as free beings. Despair is in fact the door to liberty. Grimault explains Kierkegaard's position thus: "It was healthy to despair for oneself and, from the depths of this despair, to choose oneself in one's eternal value. In fact, I can then choose only in an ideal sense. And what is that ideal? It is myself in my eternal value. And this person in his eternal value is liberty. Through this choice, man abandons the aesthetic for the ethical."[9]

Given our rootedness in the world, spiritual peace is not accessible; only a certain serenity is within reach. The reason for that serenity has a name. It is God.

How tormented, how psychologically out of balance, seem the works and the position of Kierkegaard today. We must admit that he was raised in a very peculiar atmosphere, marked not only by the stern Christianity of that period but also by a family drama concerning his father, which he always kept secret.

JOY COUNTERING THE ABSURD

How much more harmonious and yet related overall to that of Kierkegaard seems the solution to the problem of the absurd proposed by Henri Bergson, the philosopher of duration.

Bergson found himself in agreement with Kierkegaard that a definitive reason for our anxiety is the cognizance of our inevitable death. Animals do not know that they must die, he noted. Humans know it. Seen from this point of view, religion is a natural defensive reaction against the intellectual representation of the inevitability of death (*Les Deux Sources de la morale et de la religion* [The two sources of morality and religion]). And again: "Anxiety is born when man attempts to foresee the unforeseeable. Early man had little grasp of nature and secondary causes. In this state, mythical causes proliferated, on which Lévy-Bruhl, among the first, insisted."[10]

At this point Bergson remarks with good sense that the *primitives* depended on an implicit regularity of causes and effects: "We are not told that the primitive, seeing the wind bend a tree, the wave rolling the pebbles, even his foot raising dust, raises the specter of something other than what we call mechanical causality. The occult concerns precisely that which has *a human significance.* If one of these events affects man, it must, thus, have some kind of intention. In this sense, there is no such thing as chance."[11]

Like the preceding philosophers, Bergson allows for the possibility of a fundamental disorder in the world. What distinguishes him, where he differs fundamentally from Kierkegaard, is that he allows for the possible relationship between this disorder and a direction in evolution and the world's becoming. This is the fundamental theme of *L'Évolution créatrice* (Creative evolution). In particular, he sees in the evolution of species, through Darwin's proposed chance variations, something that reveals the existence of a direction. As an example he cites the fact that a single organ, such as the eye, could appear in different living orders. The materialists pretend, he argues, that natural selection is responsible, in each and every case, for the appearance of the eye from primitive photosensitive pigments. It is already hard to believe that such a complex type of organ could have appeared in even a single species through a simple interplay of chance variations and natural selection. How then is it possible to explain on such a basis that the eye of a mollusk resembles that of a vertebrate? That the eye of a modern mammal resembles that of a modern bird?

> There is, in vision, something *more* than the component cells of the eye and of their reciprocal coordination. In this sense, neither the mechanism nor finalism go as far as they should. But, in another sense, mechanism and finalism both go too far, since they attribute to nature the most difficult labors of Hercules by pretend-

> ing that it could marshal an infinity of infinitely complicated elements in order to create the simple fact of vision, when nature has, in fact, no more trouble making an eye than I have raising my hand. . . . But this is what we have difficulty understanding, because we cannot keep ourselves from seeing organization as *fabrication.* It is, however, one thing to fabricate and another thing entirely to organize.[12]

In this way Bergson insists that the appearance of an organ must be understood not as a fabrication but as an organization. A fabrication requires that preexisting elements be assembled, as in assembling transistors and soldering them together to create a radio. But the creation of the eye from a single primitive photosensitive cell indicates that that cell itself possesses an organizing power that ensures that it will attract other photosensitive cells, which will then intrinsically organize themselves into an eye. In this way we can explain how the eye of the mollusk can resemble the eye of the vertebrate. But this also presupposes an incredibly powerful organizing potential in the living being. Bergson goes so far as to think that this organizing potential in life allows it continuously to defy the second law of thermodynamics. Today we know that the attribution of such vitalistic qualities, transcending the laws of thermodynamics, was a grave error.

Bergson did not stop there. He distinguished, in addition to the organizing power of life, a third level of transcendence, one belonging only to the human mind and manifested not by intelligence but by intuition. In this radical dualism, where does the problem of anxiety stand? As we have seen, for Bergson anxiety comes from our foreknowledge of death. But in *L'Énergie spirituelle* (Spiritual energy [1919]), Bergson indicated how, contra Kierkegaard, existential anxiety could be transformed into a profound joy:

> The philosophers who speculated on the significance of life and on the destiny of man did not pay enough attention to the lesson that nature took the trouble to teach us about herself. She tells us, with a specific sign, that our destination has been attained. That sign is joy. . . . If we pay attention to this signal, we find that wherever there is joy there is creation. The richer the creation, the more profound the joy. The mother who looks at her child is joyous, because she knows she has created him, physically and morally. The businessman who develops his business, the factory foreman who sees his industry prosper, is he joyous because of the money he will gain and the fame he will acquire? Wealth and notoriety obviously enter into his satisfaction, but they bring him pleasures rather than joy, and his true joy comes from having built an enterprise that works, from having brought something to life. Take exceptional joys. Those of the artist who realizes his thoughts, those of the thinker who discovers or invents. You are told that these men work for glory, and that they take their deepest joy in the admiration they inspire. Great mistake! Praise and honors are desired only in proportion to the degree of one's uncertainty about having succeeded. There is modesty underlying vanity; it

is in order to reassure oneself that one seeks approval, and it is in order to support the perhaps insufficient nature of one's work that one wishes to surround it with the admiration of other men, like putting the premature child into cotton wool. But he who is sure, absolutely sure, of having produced a viable and durable work has no need of praise and feels himself above glory because he is a creator, because he knows it, and because the joy he feels for it is a divine joy. If thus in all domains, the triumph of life is creation, should we not suppose that human life has its reason for being in a creation that can, differently from that of the artist and the thinker, be pursued at all moments by all people? The creation of self by self? The aggrandizement of personality by an effort that makes a lot from a little, something from nothing, and adds ceaselessly to the richness of the world?[13]

LIVING WITH CHANCE

Needless to say, science cannot force us to choose between the profane or sacred despairs of some and the spiritual joy of others. But when, in fields as diverse as physics and the biological sciences, it finds hints that the existence of a true chance is more and more likely, although it is veiled behind the appearances that are the only things to which we have access, science asks us to take this perspective into account. Contrary to what one might think, this is a relatively new perspective, since, as Bergson reminded us, the primitive way of thinking, quick to resort to magic, leaves no place for chance. If chance exists, then I am the child of chance, and even my death will change little in that reality. Alain Badiou commented on this idea, that chance interposes itself between being and knowledge.[14] But this is not enough of a description, since the sciences suggest that chance exists in itself. It is for me to choose whether I can live with it in stoicism, in despair, or in joy, knowing that no power in the world, no laboratory apprentice sorcerer in the expanding cognitive sciences, can dictate this choice. To those who choose joy, however, science offers this important comfort: that insofar as chance has an intrinsic creative power, their choice perhaps not only is dictated by instinct or the principle of pleasure but corresponds to a coherent vision of the world.

CONCLUSION

THERE IS NO CHANCE WITHOUT LAW

What is marvelous about nature, exclaimed Einstein, is that it is intelligible. And it is true that nature is largely intelligible. The entire sphere of research into laws, which amounts to a summarizing of nature, has shown itself to be both possible and potent. But Einstein also said that God is subtle, not spiteful. Subtle he is indeed. To all appearances, he chose to write nature using, so to speak, two musical registers. One of them covers the level of deterministic designs, which provides us with a great deal of our comprehension of the universe but which excludes any explanations for becoming, for surprise, for the new. And then there is the register of events without cause, of chance. Chance allows us to account for the evolution of the universe, for the things that surround us, for life. Nature sprawls over the two banks of a great river, or over two planes on which our intellectual microscopes cannot focus at the same time.

As I have already discussed, the search for a positive definition of chance has always failed. The sphere of chance can be defined only negatively, as that of events whose description allows no reductive synthesis. This does not mean that chance does not provide clues for us to find. On the contrary, the entire development of the calculus of probabilities, statistical mechanics, and thermodynamics up through quantum mechanics demonstrates that *there is no chance without law.*

Thus, chance is not the enemy of rationality that certain people have thought it to be. Its rationality, however, is different from deterministic rationality. And to paraphrase and expand on Einstein's statement, it could be said that what is most marvelous in nature is that even the part of it that escapes causality does not entirely escape our minds' abilities to define it, to enumerate its laws. In fact, it is through the laws of chance that we can trace its existence.[1]

We must take account here of the true import of these laws, which treat more than just the results of a large number of effectively performed trials. The difficulties with the classical interpretation of probabilities, especially with negligible probabilities and the meaning to give to probabilities of unique events, were resolved essentially by the neoclassical interpretation of probabilities in terms of measures, in the mathematical sense of the term. Contrary to the classical frequentist interpretation, the modern interpretation of probabilities in the light of the theory of measures allows us to replace the limits for an infinite number of trials with measures that can apply to singular series. It also allows attributing a probability to a unique event, as long as there exists an unlimited group of virtual series that might complete it.[2] Karl Popper even thought that, if the interpretation he proposed for probabilities (in terms of *propensities*) were accepted, what he called the fundamental problem of chance could be resolved and easily interpreted in the framework of Kolmogorov's system of axioms. "Joined to the propensionist interpretation," he explained,

> this latter is able to *explain why* the sequences of independent events . . . have the strange behavior that is observed—*why* they behave as von Mises collectives, *why* their frequencies exhibit a tendency to converge; *why* they have, at the same time, an aleatory character, so that (almost) all the game systems fail. The fundamental problem of the theory of chance is resolved once we are able to understand this strange behavior, joining regularity to irregularity—behavior that in one sense each of us expects from events governed by chance but that constitutes, no less, a considerable problem for whoever thinks seriously about it.[3]

Yes, but what is the reality of these propensities? Are they properties attached to the events themselves, one by one, individually? Popper answers in the negative. Propensities, he explains, are relational properties.

> The tendency of statistical averages to remain stable, if the conditions are also stable, constitutes one of the most remarkable characteristics of our universe. We cannot account for this, it seems to me, except with the propensionist theory, in other words the idea that there exist possibilities that are *more than simple possibilities.* These are tendencies or propensities to realize themselves. Inherent, in differing degrees, to all possibilities, they are like the forces which ensure the stability of frequencies. . . . We suppose that propensities are not simple possibilities, but that they have a physical reality. They are as "real" as forces or force fields. And *vice versa:* forces are propensities for making a body move, accelerate, and fields of forces are propensities distributed over a given region of space. . . . The fields of forces are fields of propensities. They are real, they exist.[4]

This is a thought that he completes elsewhere. "A Newtonian force is not the property of an object, but the relational property of at least two objects; and the forces that result, effectively, in a given material system always constitute a property of the system as a whole. The force, like the propensity, is a relational concept."[5]

We are not obliged to adopt Karl Popper's point of view in its entirety, of course, and there are certainly epistemologists who criticize him. But we must still admit to a convergence between the modern approach to probability and his point of view, which agrees also with the interpretation we might be tempted to give of the properties of natural chance, when its teleonomic power is seen.

TELEONOMIC CHANCE

By definition, chance-produced events have no antecedents. No hand, no providence external to those events directs them. Otherwise they would no longer be random.

By its very nature, mathematical chance can lead nowhere. What it does today, it can undo tomorrow, in total blindness. It is absurdity itself. But chance, as we have seen it incarnated in nature, from cosmogony to biological evolution, decidedly avoids that relegation to the absurd. There is thus a difference between physical chance and mathematical chance. The difference is this: mathematical chance is not a builder. Physical chance constructs, even if it is without a final goal. No one directs and no one uses chance. In this sense it contradicts the old teleonomic designs. The finality of physical chance is inherent in chance itself. And what is remarkable is that this leads to the organization of the original void into majestic structures deployed in the cosmos, with the spark of life and thought, even if none of these forms of organization is a priori necessary. "For Monod," Hubert Reeves commented, "the essential role of chance in biological evolution proves the absence of an intention in nature. It renounces as illusory, therefore, the ancient alliance of man with the Universe. Man is an accident along the way, in an empty and cold cosmos. He is a child of chance. Certainly. But of 'a bridled chance.' Let us salute the nature that brought 'chance' to its knees in order to enlist it as an admirable ally."[6]

Karl Popper developed this idea that natural chance is endowed with a sort of "*a priori* cognition." "Kant," he explained, "thought *a priori* cognition was a cognition that we possess *prior* to sensory observation; and *a posteriori* knowledge was a cognition we possessed *after* observation. . . . I will thus use the term '*a priori*' in order to characterize the type of (fallible) cognition that an organism possesses *prior to sensory experience,* or more generally, as *innate* knowledge." And Popper gave, as an example of such a cognition, the invention of the eye, which had already drawn Bergson's attention. "The invention of the eye is thus the invention of a new theoretical *a priori* cognition, of an adaptation to the environment. It was from the start an adaptation to the 'long-term' structure of the environment: to the existence of solar light, potentially usable as food. The eye thus incorporates an understanding of this environmental structure. This is theoretical knowledge with a high degree of universality, almost in the same vein as the Kantian awareness of space and time."[7]

DETERMINISTIC CHAOS: AN IMPOSTOR

During these past few years we have witnessed the prodigious development of the science of disorder, in the form of the theory of deterministic chaos. The public's infatuation with this new approach to often complex scientific problems has been on a par with its liking for the pretty pictures that today's most powerful computers are giving us of fractals and strange attractors. An entire field of the science of complex systems (e.g., turbulence), previously thought to be just so many examples of the processes of chance (or "stochastic processes"), has thus suddenly become accessible by scientific methods. Events that appeared to be random have been shown to be perfectly determined phenomena, but phenomena whose continuous and deterministic trajectory can be described only in a mathematical space endowed with a higher dimensionality than that of the manifold space in which we had until then been trying to represent them.

All right. But we must not allow ourselves to be fooled and go about proclaiming that since the theory of deterministic chaos exists, the problem of chance has been resolved or cleared from the field of fundamental questions. Deterministic chaos is a perfect example, once more, of classic and legitimate research procedures, which consist of pushing back the limits of chance as far as possible. In describing these phenomena, we can only regret that an unfortunate choice usurped the beautiful name Plato had coined for the fundamental and original substance of the universe. The science of chaos, as it is described in these new treatises, is not at all the science of chance but that of its counterfeit in the universe. The existence of counterfeits has never cleared away the problem of the existence, real or not, of the original.

A SEPARATION THAT GIVES MEANING

If the original does truly exist, it is, as I have tried to show throughout this book, lurking behind a curtain, a separation. The idea of cutting short all metaphysical speculation on the origin of probability to assert its existence and verify its practical consequences is certainly due in large part to Darwin.

In the theory he constructed, Darwin brought in the chance of variations as it appeared to naturalists. In doing so, he not only transposed but *proposed* chance as the foundation of biological evolution. The word that best summarizes the role assigned to chance in his thinking and in his work is perhaps *opportunism.* The passage from variations as they are observed in the practice of artificial selection to the chance envisioned for natural selection is one of the points that defines the difference between Darwin and Alfred Russel Wallace, the other inventor of biological evolution. In fact, it is precisely on the question of chance that the works of these two pioneers of modern evolutionism diverge. Wallace considered it possible that the direction of evolution is entirely determined by the mechanisms of selection that apply to an entire population. Consequently, his evolutionism could very well have been reconciled with natural theology. He did not consider that a variation could create the conditions for a new selection and that this selection would be subject to the constraints applying to variations. For Darwin, on the contrary, the one-way linkage between variations and selection could lead only to the absence of direction, to the absence of finality. The *opportunism* of living chance can be applied entirely blindly and can determine the destiny of evolution.

Reluctantly Darwin made a significant change in his position on the question of addressing finality. Before him, the question amounted to a choice between efficient causes and final causes. With him, this question became that of an irreconcilable separation between causes and effects, between chance and evolution. In fact, with Darwin, what was important was that the "end" is not the cause of the effects that attain that end. It is exactly this that causes the difficulty in his reconciliation with natural theology, at least as it was understood in the nineteenth century.

The introduction of this theoretical separation can, thinks Lenay, allow us to understand that chance is neither "subjective" nor "objective." It is "theoretical," something that has meaning only within the framework of a constructed and thus historical scientific theory.[8] But where should the theory itself be placed? On the side of nature, as Einstein seemed to want it with his theoretical realism, or on the side of epistemology, as Niels Bohr doubtless would have liked it?

The notion of rupture associated with chance, the basis of a new and fundamentally different quality of reality (relative to what it would be without it), obviously has a metaphysical connotation. It intersects the classical distinction between the probability of ignorance and intrinsic probability. As Prigogine and Stengers rightly remind us,[9] what I will call the problem of the ontological rupture of chance bears a certain resemblance to the old difficulty, already raised by Leibniz in the seventeenth century, of going from the limits of an understanding of finite precision to an infinite knowledge that can be expressed neither in words nor in numbers and that Leibniz thought to be reserved for God.

But what Darwin did not foresee was that the problem of finality is not radically excluded by the discovery of the chasm separating physical chance; it is only displaced. The question that must now be asked is this: is there in physical chance something more than its mathematical definition describes? Remember that the mathematical definition of chance is summarized by saying that chance does not allow for any summary. If physical chance exists, it must conform to this requirement, but it need not be limited to this. It could be that physical chance is also the location of a creative urge, of a motor act that never stops going from the state of virtuality to the state of actuality, of "propensities" that never stop becoming probable events.

THE TEACHINGS OF QUANTUM MECHANICS

This all sounds familiar by now. Quantum mechanics raises the problem of Schrödinger's cat, and it seems that to resolve it, we must accept that at one level or another there exists a separation between the microscopic world and the macroscopic world, that of phenomena as we perceive them. But as is well known, we do not know the precise location of this separation. It is even, in this case, in part arbitrary.

Among the authors who came the closest to conceptualizing chance as I have been explaining it, Karl Popper is the one I most want to retain. In a seductive analysis of this author's thinking, after having remarked that in the orthodox interpretation of quantum mechanics "irreversibility and probability appear to be relative to the conditions under which the quantum world can be observed," Prigogine and Stengers note that Popper was right to contend that there is no reason to identify the "dream" of a return to realism with that of a return to determinism.[10]

Quantum mechanics has revealed the need to return to a radical, essential vision of chance. "What is important and new, when it comes to quantum mechanics," notes Patrick Suppes, "is not so much the negation of determinism, which I accept as a finding of current experimentation (many ordinary experiments argue against determinism; that which induced us to see the world in a deterministic fashion was the seduction of the divine music of celestial mechanics composed by Newton and Laplace). But the negation of stochastic causality [i.e., the impossibility of appealing to hidden variables] constitutes an otherwise radical housecleaning."[11]

We find ourselves thus pushed to propose the existence of intrinsic probabilities. Popper, as we know, baptized these as *propensities.* If they unfortunately appear unable entirely to resolve the epistemological problems raised by quantum mechanics, as Bernard d'Espagnat's analysis suggests,[12] they nevertheless allow some to be resolved and allow us to find ourselves closer to a satisfactory understanding of the nature of nature.[13]

PROOF THROUGH THE CONCEPTUAL POWER OF CHANCE

Another of the essential reasons that encourage us to look favorably on the realism of chance is the practical utility of this belief. The concepts that the sciences forge of nature are at first only useful words because they help us to comprehend nature. Eventually a more detailed examination can give us additional reasons to believe that the potency of concepts introduced in this way is not only semantic but derives from a sort of ontological existence. Think, for example, of the quiet introduction of the concept of electromagnetic fields into physical theory and the epistemological place it occupies today.

The concept of chance is equally subject to this law. I believe in chance because, with this concept as a tool, it is easier for me to understand the difference between heat and kinetic energy. I believe in chance because with this concept it is easier for me to understand temporality, temporal irreversibility, not as a sensory illusion or as a filter imposed by the functioning of my brain but as a real property of nature. I believe in chance because it helps me to see the power of life, the breath that animates matter and drives it into forming the tree of life.

Some people might say that all I am doing here is transposing the ancient theological proofs of the existence of God, whose weaknesses have been exposed. I do not share the enthusiasm of some scientists and philosophers for the hints of agreement they find in an interpretation of modern cosmological findings within the framework of an "anthropic principle." I think they forget too easily the myopia that our nature as creatures of concrete existence immersed in a particular world induces in us: we can see only the local well in which our nature has trapped us and are incapable of conceiving of the other "local wells" that represent other possible universes. However, I do accept the just evoked parallelism between the presented justification for the realism of chance and Saint Anselm's proof of the existence of God. Science, however, has made some progress since this Catholic priest came up with his argument. Simply conceiving of the perfection of God is probably not proof of his existence if he does not agree to speak to us. Chance may perhaps be God's voice. In any case, that great devil Bergson was right when he wrote: "Philosophy is not only a returning of the being to itself. . . . It is also an effort to understand true evolutionism, and from there the true extension of science."[14]

CHANCE DOES NOT THREATEN THE SCIENTIFIC IDEAL

I have given my reasons for leaning toward an objective interpretation of chance, even if it requires accepting the existence of a radical rupture, of a clear distinction between an ontological *this side of* and a phenomenological *beyond* resembling the Kantian distinction between the world in itself and phenomena. This interpretation alone, it seems to me, renders temporal irreversibility understandable: I think that on this point modern science has upheld, rather than dismissed, Charles Sanders Peirce's arguments. It remains true, however, that such a rupture marks a boundary to the reductionist ideal of science: reductionist approaches, which try to explain observations at a given scale by determinations at a lesser scale (e.g., the chemical properties of bodies by the atomic structure of their molecules), cannot be endlessly pursued. There comes a time when the soil erodes beneath their steps.

But I do not believe that this must be a cause for regret or scandal, neither for René Thom nor for anyone else. I embrace the scientific ideal—that is, the attempt to push our knowledge *as far as possible* using scientific tools. In contrast to René Thom, however, who thought it indispensable to assert that our knowledge is potentially unlimited (which, after all, given what we know that we do not know, is a metaphysical attitude), I accept that a limit does exist. The essential point that can reconcile us is that no one knows where to place that limit. For this reason alone there is an unlimited field remaining for scientific research. Therefore I do not see how, given these conditions, supporters of a fundamental chance can be accused of betraying the scientific ideal.

I am not trying here to adopt a positivist and dualist vision of chance. Chance introduces us to the veiled mysteries of true reality, but as we study it, chance does not necessarily force us to abandon the prudently realistic attitude that has always been and still is the scientific ideal. The epistemological program being presented here is obviously not compatible with Costa de Beauregard's vision, if we accept Campbell's summary of his viewpoint: "Entropy is an aspect of probability, and probability operates as the hinge between matter and mind, where one is knotted to the other, and reacts on the other."[15] At this level, it seems to me, the problems of information and of the nature of probability have been arbitrarily confused. The principle of parsimony strongly urges me to beg the cyberneticians and other communication theorists not to lead us to introducing under the term *information* some quantity that has a new, nonreducible quality. And I do not believe that, for it truly to exist, chance or probability requires this sort of rampant dualism. On the contrary, we can preserve the unirealistic scientific ideal and still believe in chance, as long as we specify that we do not know where to place that rupture between causeless events and events that obey laws. I believe in chance, but I swear that I shall wholeheartedly pursue the endless effort to push back the horizons of the random.

It might perhaps be appropriate to reiterate now that I am not blazing any new trails by arguing for the reality of chance. The number of those defending similar points of view is constantly growing among scientists, while those defending an absolute determinism often find themselves fighting a defensive battle. But it is certainly appropriate to recall that scientific revolutions, dear to Thomas Kuhn, or changes of paradigms, to speak as Gerald Holton, do not occur quickly and are not the work of a solitary researcher. "A scientific revolution," observes Alan Chalmers, "means abandoning one paradigm and adopting a new one, not by one person, but by the entire interested scientific community. As more are converted, for whatever reasons, to the new paradigm, there is an 'increasing change in the distribution of professional thought' (according to the formula used by Kuhn in *The Structure of Scientific Revolutions*)."[16] We are now possibly and even probably present at one of these turning points, at the end phase of a scientific revolution so courageously begun in 1859 by two pioneers, James Clerk Maxwell and Charles Darwin. The change should allow science to regain the broader horizons that were constricted by the classical vision and to reinvigorate and launch new research in many fields. Some may find the easy trail of certainties to be lost, but others, such as Ilya Prigogine, will find there the possibility for a new alliance between humankind and nature.

APPENDIX 1

THE LAW OF LARGE NUMBERS

BERNOULLI'S THEOREM

Let A be an event that may result from chance, and consider a series of n identical and independent random trials during which event A may or may not take place. Let n_A be the number of trials during which A is observed, P(A) be the probability of A, and ϵ and η be two arbitrarily small numbers. The frequency of observing A is defined as n_A/n. There exists a number $n_o(\epsilon, \eta)$ such that

$$\text{If } n \geq n_o \text{ then Prob } [\ |\ (n_A/n) - P(A)\ | \geq \varepsilon\ ] \leq \eta$$

Proponents of an objective interpretation of probability hold that the measurement of frequency is the only way to evaluate probability concretely. As rightly noted by Bonitzer,[1] however, the theorem opens the door to an infinite regression, since the divergence between probability and frequency is subject to a second-order probability (the probability of a given divergence) that must in turn be evaluated by frequency method, and so on.

For proponents of this school of thought, this infinite regression is an embarassment that must sooner or later be overcome.

To get around it, supporters of a subjective interpretation of probability proposed treating the two concepts of probability and frequency as equally accessible to direct empirical measurement (probability being measured by means of psychological testing). If this is the case, however, why should the two concepts tend to converge when the series of tests is extended? They converge because of a need for logical coherence, reply Bruno de Finetti and Leonard Savage, modern statisticians who support this point of view. Experience teaches. But then the distinction between the two schools' positions tends to vanish. The brain, which evaluates probabilities, is forced to apply itself to learning and to rank itself with the verdict of the frequencies.

CHEBYCHEV'S LAW OF LARGE NUMBERS

Let $X_1, X_2, \ldots, X_n$ be a series of independent random variables obeying the same distribution of probability and having a common mathematical expectation, $E(X) = m$. There exists a number n_o such that, for every set of arbitrarily small numbers $\epsilon, \eta > 0$,

$\bar{X}_n = \frac{1}{n} \Sigma_k X_k$ satisfies

Prob $[\ |\bar{X}_n - m| \geq \varepsilon\ ] \leq \eta$ for any $n \geq n_o$.

One can see that Bernoulli's theorem is a particular case of Chebychev's law, when the random variable X takes the value 1 or 0 according to whether event A is realized.

Chebychev's law was one of the first examples of the marked tendency, at the end of the nineteenth century, to restore mathematical expectancy's preeminence over the concept of probability. Despite the generalization it introduces, the new law of large numbers encountered the same difficulties of interpretation as did Bernoulli's.

BOREL'S "STRONG" LAW OF LARGE NUMBERS

In using the concept of almost certain convergence, the theorem demonstrated by Borel in 1909 reads more simply than the preceding laws, since it no longer makes finite limits intervene for the probability of divergence. It simply stipulates that for a series of independent trials on a random variable with a finite expectation m, the average of the observed results $\bar{X}_n$ tends almost certainly toward this mathematical expectation:

$$\bar{X}_n \underset{\text{p.s.}}{\rightarrow} m.$$

The limit of a *null* probability (when $n \rightarrow \infty$) is thus substituted for the very small probabilities considered by Bernoulli and Chebychev. The obvious advantage is that this approach eliminates the logical problem linked to the evaluation of the probability of divergences. There still remains, however, an infinite regression. In effect, to say that $\bar{X}$ almost certainly tends toward m is to say that, for an indefinitely prolonged series of trials, the limit of the average $\bar{X}_n$ is equal to the mathematical expectation m with a probability = 1. This in turn does not exclude the possibility that the average does not converge toward the expectation in some series of trials, provided that such extraordinary series are so rare as to constitute "negligible events." In other words, in an infinite succession of an infinite series of trials, the average does not differ from the mathematical expectation more than a finite number of times. The probability of these extraordinary successions is indeed null in a relative value. But to ensure that the series of trials being attempted is not one of these extraordinary series, it must be worked out . . . an infinite number of times.

APPENDIX 2
BAYES'S THEOREM

Bayes's theorem relies on the law of the composition of random events, which states that the probability of realization of two events A and B is equal to the product of the probability that A will occur and the probability that B will occur given that event A has already occurred. This intuitive rule is easily verified thanks to a visual representation of the ensembles of the possible events. In the notation of the probability theory, it is written:

$$\mathrm{Pr}(A \cap B) = \mathrm{Pr}(A)\ \mathrm{Pr}(B/A),$$

where Pr(B/A) represents the *conditional probability* that B will occur given that A has already taken place.[1]

Naturally, the same probability can also be stated as a function of the probability that B will be realized:

$$\mathrm{Pr}(A \cap B) = \mathrm{Pr}(B)\ \mathrm{Pr}(A/B).$$

Let A be a physical event whose realization may be the consequence of any one of the causes (each exclusive from the others) $B_1, B_2, \ldots, B_n$, with no certainty as to which of the causes is responsible for A. Let us in addition call $\mathrm{Pr}(B_1), \mathrm{Pr}(B_2), \ldots, \mathrm{Pr}(B_n)$ the a priori probabilities that causes $B_1, \ldots, B_n$ are present independently of A. The probability that B_i and A are connected—that is, the probability that A is due to B_i—is written

$$\mathrm{Pr}(B_i \cap A) = \mathrm{Pr}(B_i)\ \mathrm{Pr}(A/B_i) = \mathrm{Pr}(A)\ \mathrm{Pr}(B_i/A).$$

Thus the probability of B_i is

$$\mathrm{Pr}(B_i/A) = \mathrm{Pr}(B_i)\ \mathrm{Pr}(A/B_i)/\mathrm{Pr}(A).$$

In the same way, the probability that another cause B_k produced A is

$$\mathrm{Pr}(B_k/A) = \mathrm{Pr}(B_k)\ \mathrm{Pr}(A/B_k)/\mathrm{Pr}(A).$$

Finally, the ratio of the respective probabilities of causes *i* and *k* is

$$\frac{\mathrm{Pr}(B_i/A)}{\mathrm{Pr}(B_k/A)} = \frac{\mathrm{Pr}(B_i)\ \mathrm{Pr}(A/B_i)}{\mathrm{Pr}(B_k)\ \mathrm{Pr}(A/B_k)}.$$

We can see that the probabilities of the *causes given the effects* are not proportional to the probabilities of the *effects given the causes* except in the particular case where the different causes have the same likelihood at the beginning, that is, the same a priori probability. This is the essence of Bayes's theorem.

APPENDIX 3

THE PRINCIPLES OF QUANTUM MECHANICS

Quantum mechanics predicts, with fabulous precision, the results of our observations of the behavior of atomic or nuclear systems and the properties of materials. Since its creation in the thirties, it has accumulated success on success, without yet failing once. The general principles of this theory may be presented in six points.[1]

1. THE IDEAL SPACE OF QUANTUM MECHANICS

Quantum mechanics is deliberately indeterministic. It substitutes a conceptual space, that of "state vectors," for real space as an area for the application of pure causality. A given microscopic system is conventionally represented by a mathematical object, its state vector, which since Dirac physicists have represented by a "ket": $|>$.

Kets are distinguished from one another by the values that characterize the system. If a particle has a defined momentum p, for example, it is represented by the ket $|p> = |p_x,p_y,p_z>$. If it has a defined position r, it is represented by the ket $|x,y,z>$.

2. PHYSICAL QUANTITIES IN QUANTUM MECHANICS

The name *quantum mechanics* indicates that in this theory, the physical values attributed to a microscopic system (e.g., position, energy, momentum) cannot vary arbitrarily. Some values are allowed, but others are forbidden. Thus, for example, the energy of an electron in a hydrogen atom is *quantized:* there exist a series of possible values $E_0, E_1, \ldots, E_n, \ldots$, but intermediate values are impossible. The smallest of the authorized values for energy, E_0, represents a limit below which the energy of the electron cannot descend. Thus, this limit actually ensures the atom's stability.

Formally, the authorized values of physical quantities are obtained in the following way. Each quantity that characterizes a system is represented by an "operator" that acts on the kets by transforming them into one another. The operator "energy," E, for example, transforms the ket $|x,y,z>$ into another ket that I will conventionally designate as $|\Phi>$. This transformation is written as follows:

$$\mathrm{E}\,|x,y,z> = |\Phi>.$$

The operator "projection onto Ox of momentum," P_x, transforms the ket $|p>$ into itself, to within a multiplicative constant, which is exactly the momentum p_x of the particle:

$$P_x|p_x,p_y,p_z> = p_x|p_x,p_y,p_z>.$$

More generally, all the quantities characterizing a quantum system can take only values for which an equation similar to the preceding equation (called the "eigenvalue equation") is satisfied. The quantity A, for example, can take only the values $a_0, a_1, \ldots, a_n, \ldots,$ for which there exist kets $|a_i>$ such as:

$$A|a_i> = a_i|a_i>.$$

The kets $|a_i>$ that satisfy the previous equation are called "eigenvectors" of operator A, and the multiplicative constant a_i is called the "corresponding eigenvalue."

3. THE PRINCIPLE OF SUPERPOSITION

In general, a given state vector $|\Phi>$ is not identifiable as one of the eigenvectors $|a_i>$ of operator A, but it can always be written as a superposition—that is, a sum—of eigenvectors:

$$|\Phi> = \Sigma_i\alpha_i|a_i>,$$

where the α_i can be real or complex numbers (i.e., in the form $a + ib$, where $i = \sqrt{-1}$).

4. INTERPRETATION OF QUANTUM MECHANICS

If the ket representing the quantum system under consideration is an eigenvector of operator A representing the quantity A—for example, $|a_k>$—the measurement of A on the system will certainly result in the value a_k.

If the ket representing the quantum system is a superposition of eigenvectors of operator A—for example, $|\Phi> = \Sigma\alpha_i|a_i>$—the measurement of A on the system can result in any one of the eigenvalues contained in this superposition. The probability p_i of getting result a_i is given by the relationship $p_i = \alpha_i\alpha_i^*$, where α_i^* is the conjugate of α_i (if $\alpha_i = a + ib$, then $\alpha_i^* = a - ib$, and $p_i = a^2 + b^2$ is always a real positive number, as is every probability).

5. THE PRINCIPLE OF THE COLLAPSE OF THE WAVE PACKET

If the measurement of the quantity A on a quantum system has resulted in the value a_i, the ket that represents the state of that system just after measurement is the corresponding ket $|a_i>$, whatever might have been the ket representing the system before measurement. (The name given to this principle refers to the old wave mechanics, for which the wave function associated with the system, just like its representative ket, changes abruptly during the measurement.)

6. SCHRÖDINGER'S EQUATION

When no measurement is taken, and the quantum system is left alone, the ket $|\Phi>$ that represents it evolves in a *causal fashion,* in conformity with the first-order differential equation relative to time:

$$i\hbar \frac{d|\Phi>}{dt} = H|\Phi>,$$

where H is the operator corresponding to the total energy of the system (this definition coincides with Schrödinger's equation written for wave functions in wave mechanics).

CONSEQUENCES

The axioms of quantum mechanics have consequences surprising to those trained in the spirit of classical mechanics. Among the most important of these consequences is *the indeterminism of our relationships with nature,* since it is obviously impossible to follow, in a strict fashion, the causal history of a microscopic system: the measurements change the ket representing the system in a way that is predictable only in terms of probabilities. Another important consequence is the *indiscernability of identical particles,* since we cannot follow the trajectories of each of them except in the instants in which we measure them. Thus, if two identical particles collide between two of those instants, it is impossible to know which of the particles identified after the collision corresponds to a given particle in the initial state. The impossibility of describing the history of identical microscopic systems obliges us to admit that they share their individuality. Finally, we can demonstrate that *Heisenberg's indeterminacy relationships,* which determine the limits of the precision with which one can simultaneously assign to an elementary particle both position and momentum (or other pairs of so-called *complementary* quantities), are direct consequences of the preceding axioms.

NOTES

INTRODUCTION

1. R. Lestienne, *The Children of Time: Causality, Entropy, Becoming* (Urbana: University of Illinois Press, 1995).

2. Quoted in C. Lenay, "Enquête sur le hasard dans les grandes théories biologiques de la seconde moitié du XIXe siècle" (Enquiry on chance in the great biological theories of the second half of the nineteenth century), Ph.D. diss., University of Paris-I, 1989, p. 127.

3. M. Kimura, *The Neutral Theory of Molecular Evolution* (New York: Cambridge University Press, 1983).

4. D. Brooks and E. Wiley, *Evolution as Entropy,* rev. ed. (Chicago: University of Chicago Press, 1988).

5. S. Gould, *Wonderful Life* (New York: Norton, 1989). See especially the figure reprinted in this book on page 63.

CHAPTER 1: A BRIEF HISTORY OF THE CONCEPT OF CHANCE

1. See W. Heisenberg, *Physique et philosophie,* rev. ed. (Paris: Albin Michel, 1971), 240; English-language version, *Physics and Philosophy* (Chicago: Encyclopedia Britannica, 1990).

2. This preoccupation with fairness can be better understood if we keep in mind the first softenings of the Catholic church's hitherto rigid doctrine toward moneylenders. In 1645 the Jesuits obtained, for the first time, a dispensation from the sacred congregation of the faith for Chinese converts who practiced lending at interest (30 percent!), on the condition that a "just" sharing of the risks be taken into account.

3. The origin of the word *probability* seems to be linked to casuistry, according to which any sin could be absolved provided that some theologian had already taken a position of indulgence on the subject. The word was thus, at the beginning, associated with a sort of relativism. See E. Byrne, *Probability and Opinion: A Study in the Medieval Presuppositions of Post-Medieval Theories of Probability* (The Hague: Martinus Nijhoff, 1968).

4. Namely, the equal probabilities resulting from the evident symmetries in concrete situations or probabilities as numerical translations of the subjective impressions of different people regarding the chances that an event will or will not occur.

5. These first mortality tables used the findings assembled by the Protestant pastor of Breslau (Wroclaw) on request from Leibniz.

6. Intersubjectivity is the property that guarantees that a judgment of truth or falsity does not depend on the person giving the judgment. An intersubjective judgment is necessarily unanimous. Science requires that the propositions it pronounces be accepted by all (on the evidence of the observations) but does not necessarily demand that these propositions reflect the objective properties of the world. This is the debate between positivists and realists.

CHAPTER 2: 1859: THE TWOFOLD INTELLECTUAL REVOLUTION OF DARWIN AND MAXWELL

1. Starting his campaign at the beginning of May at the head of French troops, the emperor, who had promised Cavour that he would support the Piedmont against Austria in its efforts to construct a united Italy, prudently signed the preliminaries of the Villafranca peace accords on July 12.

2. In 1858 the Italian chemist Stanislao Cannizzaro clarified the idea of the molecule by using Avogadro's hypothesis (ennunciated in 1811) stating that equal volumes of gases contain the same number of molecules, whatever the nature of the gas, provided that they are at the same temperature and pressure.

3. The first specimen of such a skull found at Neanderthal, which gave its name to that variety, was discovered in 1856. Paleontologists long refused to consider Neanderthals as a variety of *Homo sapiens.*

4. E. Renan, *L'Avenir de la science* (The future of science) (Paris: Calmann Levy, 1890); this work was written in 1848 but not published until 1890.

5. P. de Laplace, *Essai philosophique sur les probabilités* (Paris: Mme Vve. Courcier, 1814); English-language version, *A Philosophical Essay on Probabilities,* trans. F. Truscott and F. Emory (New York: Dover, 1951), 3.

6. A. Cournot, *Exposition de la théorie des chances et des probabilités* (Exposition of the theory of chance and probability) (Paris: Hachette, 1843), preface.

7. He was not the first to do so. The first author to have accorded special attention to the fundamental notion of independence in the theory of probability was surely the English mathematician of French origin Abraham de Moivre, in his book *Doctrines of Chances* (1718).

8. Cournot, *Exposition de la théorie des chances,* para. 40.

9. A. Cournot, *Matérialisme, vitalisme et rationalisme* (Materialism, vitalism, and rationalism), rev. ed. (Paris: J. Vrin, 1979 [1875]).

10. See the analysis of Cournot's thought in C. Lenay, "Enquête sur le hasard dans les grandes théories biologiques de la seconde moitié du XIXe siècle" (Enquiry on chance in the great biological theories of the second half of the nineteenth century), Ph.D. diss., University of Paris-I, 1989, pp. 25–62.

11. For this subject, see R. Verderal's text on A. Cournot in *Histoire de la philosophie,* vol. 4, *La Philosophie du monde scientifique et industriel* (History of philosophy, vol. 4, The

philosophy of the scientific and industrial world), ed. F. Chatelet (Paris: Hachette, 1973), and Lenay, "Enquête sur le hasard."

12. In L. Krüger, L. Daston, and M. Heidelberg, *The Probabilistic Revolution* (Cambridge, Mass.: MIT Press, 1987), 80.

13. C. Darwin, *The Autobiography of Charles Darwin, 1809–1882, with Original Omissions Restored,* edited and with an appendix and notes by Nora Barlow (London: Collins, 1958), 122.

14. Lenay, "Enquête sur le hasard," 147.

15. Cited by R. Young, *Darwin's Metaphor: Nature's Place in Victorian Culture* (New York: Cambridge University Press, 1985), 116.

CHAPTER 3: THE BIRTH OF PROBABILITY THEORY

1. Cited by I. Hacking, *The Taming of Chance* (New York: Cambridge University Press, 1990), 95.

2. See J. Laskar and C. Froeschlé, "Le Chaos dans le système solaire" (Chaos in the solar system), *La Recherche* 232 (1991): 572.

3. I. Ekeland, *Au hasard: la chance, la science et le monde* (By chance: randomness, science, and the world) (Paris: Le Seuil, 1991), 164.

4. This number is the sum of the geometric series $2 + 4 + \ldots + 2^n$.

5. P. de Laplace, *Essai philosophique sur les probabilités* (Paris: Mme Vve. Courcier, 1814); English-language version, *A Philosophical Essay on Probabilities,* trans. F. Truscott and F. Emory (New York: Dover, 1951), 39.

6. R. von Mises, "Uber kausale und statistiche Gesetzmässigkeiten in der Physik" (On causal and statistical regularities in physics) (1930). Cited by K. Popper, *La Logique de la découverte scientifique* (Paris: Payot, 1978), 211; originally published as *The Logic of Scientific Discoveries* (London: Hutchinson, 1959).

CHAPTER 4: THE PROBABILITIES OF CAUSES

1. M. Watanabe, "Symmetry of Physical Laws, Part III: Prediction and Retrodiction," *Reviews of Modern Physics* 27 (1955): 179.

2. O. Costa de Beauregard, *Le Second Principe de la science du temps: entropie, information, irréversibilité* (The second principle of the science of time: entropy, information, irreversibility) (Paris: Le Seuil, 1963), 43–45. Watanabe's example is also cited in Costa de Beauregard.

3. A. Tversky and D. Kahneman, "Causal Schemata in Judgments under Uncertainty," in *Progress in Social Psychology,* vol. 1, ed. M. Fishbein (Hillsdale, N.J.: Erlbaum, 1980), 62. The example is also reported by G. Gigerenzer, Z. Swijtink, T. Porter, L. Daston, J. Beatty, and L. Krüger in *The Empire of Chance* (New York: Cambridge University Press, 1985), 218.

4. He could do it, for example, by minimizing the expression [Pr(G) Pr("B"/G)] + [Pr(B) Pr("G"/B)], which represents the weighted sum of the probabilities of testifying that the taxi was blue when it was in fact green and of testifying that it was green when in fact it was blue. For details, refer to M. Birnbaum, "Base Rates in Bayesian Inference:

Signal Detection Analysis of the Cab Problem," *American Journal of Psychology* 93 (1983): 85–94.

5. See especially J. Bonitzer, *Philosophie du hasard* (Philosophy of chance) (Paris: Editions Sociales, 1984).

CHAPTER 5: VARIATION AND SELECTION

1. For example, suppose that we have a pond where we are growing water lilies whose leaves grow so quickly that the leafy surface doubles each day. At the beginning, the single leaf of the young plant is minuscule, but by the end of a month the water lilies have covered the entire surface of the pond. When was half of the surface still free of coverage? The obvious answer, only one day earlier, illustrates the dangers of this type of growth.

2. C. Darwin, *The Origin of Species by Means of Natural Selection,* 6th ed. (London: J. Murray, 1888), 248.

3. C. Darwin, *The Variation of Animals and Plants under Domestication,* 2 vols. (London: J. Murray, 1885), 2:236.

4. Ibid.

5. Darwin, *The Origin of Species,* 14. Note, however, that the word *evolution* is absent from *The Origin of Species,* since it was Spencer who used it later and applied it to Darwin's theory. Darwin wanted to avoid a word that evoked the idea of progress and preferred referring to natural selection.

6. See M. Paty, "Modèles mathématiques et réalité physique" (Mathematical models and physical reality), *La Pensée* 200 (Aug. 1978): 86–101.

7. C. Peirce, "Evolutionary Love," *The Monist* 3 (Jan. 1893): 176–200; reproduced in *The Essential Peirce: Selected Philosophical Writings,* 2 vols., ed. Nathan Houser and Christian Kloesel (Bloomington: Indiana University Press, 1992), 1:352–71.

CHAPTER 6: HEREDITY'S LOTTERY

1. *Genome:* in the current sense, all the genetic material carried by an individual, including genes that are not expressed.

2. We know how this process allows the complete duplication of the original DNA. Each uncoupled branch bares a series of bases with nonsaturated hydrogen bridges, to which the complementary bases affix themselves one after another thanks to the action of an enzyme, DNA polymerase. In case of an error in copying, a correction mechanism intervenes and in most cases succeeds in reestablishing a perfect replication of the DNA.

3. C. Nägeli, *Mechanisch-physiologische Theorie des Abstammungslehere* (Mechanical-physiological theory of genealogical descent) (Leipzig: Oldenbourg, 1884).

4. M. Blanc, *Les Héritiers de Darwin* (Darwin's heirs) (Paris: Le Seuil, 1990), 205.

5. See M. Krings, A. Stone, R. Schmitz, H. Krainitzki, M. Stoneking, and S. Pääbo, "Neanderthal DNA Sequences and the Origin of Modern Humans," *Cell* 90 (1997): 19–30.

6. Is it not precisely one of the characteristics of the modern human species to stop adapting biologically once it has mastered its environment? Will future humanity in its turn see a phase of parthenogenetic reproduction in which the female will be all and the male nothing, an era where men will have to pay for their right to live through their work

for the queen, as do worker bees or warrior termites? The recent cloning of the first mammal using an adult cell as a seed reopens the question . . .

7. D. Brooks and E. Wiley, *Evolution as Entropy,* rev. ed. (Chicago: University of Chicago Press, 1988), 205.

8. Happily, more and more automated machines are being constructed and will take on the largest part of this task of analyzing DNA.

CHAPTER 7: HOW ARE SPECIES BORN?

1. N. Eldredge and S. Stanley, eds., *Living Fossils* (New York: Springer-Verlag, 1984).

2. S. Gould, *La Vie est belle: les surprises de l'évolution* (Paris: Le Seuil, 1991), 44; originally published as *Wonderful Life* (New York: Norton, 1989).

3. E. Mayr, *Evolution and the Diversity of Life* (Cambridge, Mass.: Harvard University Press, 1976), 697–704.

4. J. Melher and E. Dupoux, *Naître Humain* (To be born human) (Paris: Odile Jacob, 1990).

5. The notion of phase space is borrowed from physics, where it measures the number of distinct microphysical states compatible with the constraints defining a given macroscopic state. In the present context, phase space is the number of distinct viable forms for an organism of a given weight and of a fixed physicochemical composition.

6. D. Brooks and E. Wiley, *Evolution as Entropy,* rev. ed. (Chicago: University of Chicago Press, 1988).

CHAPTER 8: IMMUNOLOGICAL DARWINISM

1. In an active DNA molecule, the sequences of bases that are transcribed or "expressed" into messenger RNA (mRNA) are called "exons"; those that are not are called "introns." It is only the mRNA (and not the DNA) that can be considered as a "blueprint," or true code in the computer sense of the term, from which the ribosomes can translate and finally fabricate the protein corresponding to the original gene, which it does with the help of "transfer" RNA (tRNA) molecules.

2. The number of B lymphocytes present at any given moment in the human body is estimated to be about 10^{11}, divided into 10^{8} homogeneous clones of about a thousand lymphocytes each. Each clone is specialized in the production of a particular type of antibody, and the organism is constantly watched over by a comparable number of different types of antibodies.

CHAPTER 9: NEURAL DARWINISM

The title of this chapter is borrowed from Gerald Edelman's book *Neural Darwinism* (New York: Basic, 1987).

1. 0.2 microns—that is, 0.2 thousandths of a millimeter—is no more than 0.004 of the diameter of a hair.

2. The image of a tangled skein seems closer to reality than that of a pipeline insofar as the various neurons of an *assembly,* although interchangeable, clearly do not all receive

the same type of information at any given instant. On the contrary, it seems that they assume a role as precise as it is ephemeral, one after another and for short periods of time.

3. *Long-term potentiation* is the lasting, possibly permanent increase of synaptic conduction obtained under certain conditions and for certain synapses following a repeated use of these synapses to successfully transmit an action potential from one neuron to another. It probably plays a role in the mechanisms of memory. *Long-term depression* is the inverse effect, that is, a lasting diminution of synaptic efficiency. *Short-term facilitation* is its temporary and reversible increase. The excitatory synapses equipped with NMDA-type receptors (N-methyl-D-aspartate), sensitive to neurotransmitters of the amino acid type (glutamate), are synapses that are particularly plastic, subject, according to conditions, to these three types of changes.

4. C. Shatz, "Impulse Activity and the Patterning of Connections during CNS Development," *Neuron* 5 (1990): 745–56.

5. A synaptic bouton contains numerous neurotransmitter vesicles, all of the same type. The electrical impulse carried by the axon causes the potential of the synaptic bouton's membrane to change abruptly. This phenomenon can cause, but does not always lead to, the opening of a vesicle on the surface of the membrane and the release of its contents (by exocytosis) in the narrow synaptic cleft that separates the two neurons at that location (see fig. 10).

6. Specifically, in the case of an excitatory synapse, these are mostly Na^+ and Ca^{++} ions.

7. The ionic mechanisms of creation and of "active" propagation (i.e., without attenuation) of action potentials are different from those presiding over the formation of the postsynaptic potentials that may "passively" propagate along the dendrites. The action potential is characterized by the successive opening of canals for the ions Na^+ and K^+ on the axonic membrane (see J.-P. Changeux, *L'Homme neuronal* [Paris: Fayard, 1983], 103; English-language version, *Neuronal Man: The Biology of the Mind* [New York: Oxford University Press, 1986]).

8. The magnitude of the depolarization they induce on the axon's membrane can reach 100 mV, and the signal stretches from one to three milliseconds (see fig. 11).

9. Subjected to electrical excitation caused by the arrival of an action potential in the terminal fiber of an axon, each synaptic bouton has a probability p of undergoing exocytosis. The average number of vesicles liberated among the n boutons present, np, is typically on the order of at least ten.

10. B. Strehler and R. Lestienne, "Evidence on Precise Time-Coded Symbols and Memory of Patterns in Monkey Cortical Neuronal Spike Trains," *Proceedings of the National Academy of Sciences, USA* 83 (1986): 9812–16; R. Lestienne and B. Strehler, "Time Structure and Stimulus Dependence of Precisely Replicating Patterns Present in Monkey Cortical Neuronal Spike Trains," *Brain Research* 437 (1987): 214–38.

11. The original system of vision in the darkness developed by bats allows them to locate their prey to within about two centimeters, which, taking into account the speed of the propagation of (ultrasonic) echoes used, implies a precision better than 0.1 millisecond in the evaluation of the intervals between the emission of a sound and the return of the echo. See N. Suga, "Biosonar and Neural Computation in Bats," *Scientific American,*

June 1990, pp. 60–68. More recent research suggests that the discrimination power of bat sonar, and therefore the temporal discrimination power of the bat's auditory system, is even better than that. For the owl's auditory capacities, see M. Konishi, T. Takahashi, H. Wagner, W. Sullivan, and C. Carr, "Neurophysiological and Anatomical Substrates of Sound Localization in the Owl," in *Auditory Function—Neurobiological Bases of Hearing*, ed. G. Edelman, W. Gall, and W. Cowan (New York: Wiley, 1988), 721–45.

12. See M. Mézard and G. Toulouse, "Des verres de spin aux réseaux de neurones" (From spin glasses to neural networks), *La Recherche* 232 (1991): 616.

13. S. Sherman, "Development of Retinal Projections to the Cat's Lateral Geniculate Nucleus," *Trends in Neurosciences* 8 (1985): 350–55.

14. J.-P. Bourgeois, Institut Pasteur, private communication. According to Bourgeois, the relatively constant number of synaptic contacts during childhood could correspond either to a period of turnover during which the contacts that disappear are replaced by new contacts elsewhere or to a period of refinement of synaptic efficiencies without much change in the distribution of the synapses.

15. See B. Libet, "Consciousness: Conscious, Subjective Experience," in *Encyclopedia of Neuroscience*, ed. G. Adelman (Boston: Birkhaüser, 1987), 271–75.

16. J.-P. Changeux and S. Dehaene, "Neural Models of Cognitive Functions," *Cognition* 33 (1989): 63–109.

CHAPTER 10: ENTROPY, ORDER, AND CHANCE

1. There is a practical limit to exercises of this type: in particular, because of the influences of planets on one another over long periods of time, their trajectories are very sensitive to any uncertainty or to imprecision in the specification of their initial conditions. This dynamic instability makes long-term predictions (on the order of 100 million years) difficult, whether directed toward the future or toward the past (see J. Laskar and C. Froeschlé's article "Le Chaos dans le système solaire" [Chaos in the solar system], *La Recherche* 232 [1991]: 572).

2. The expression "invariance through time reversal," currently used by physicists, comes from the fact that the equations of classical mechanics or electromagnetism do not change form when one substitutes t for $-t$, where t represents time. This property is shown to be both necessary and sufficient to allow the equation to apply to both prediction or retrodiction. This symmetry through time reversal is not respected in the equation of heat diffusion. I will not discuss here the possible existence of certain phenomena of elementary particle physics that would not obey symmetry through time reversal, because I believe that they change nothing with respect to phenomenological irreversibility linked to the growth of entropy (see R. Lestienne, *The Children of Time: Causality, Entropy, Becoming* [Urbana: University of Illinois Press, 1995], 105–12).

3. The phase space of a physical system composed of N particles is the $6N$-dimensional space where each point represents a different physical state of the system. This is because the state of each of the particles that composes it is characterized by three spatial coordinates (representing its position in space) and three coordinates of momentum (representing its velocity vector).

4. The growth of entropy is measured by $\Delta S = \int dQ/T$, where dQ measures an elementary heat exchange between the system and the external environment and T is the absolute temperature at which this exchange takes place. The integral is taken from the initial state to the final state of the system, throughout an ideal reversible process.

5. Background radiation has remained identical to what it was about a half-million years after the Big Bang, except that is has uniformly cooled. Its characteristic temperature has gone from some millions of degrees to 2.7 degrees absolute. The small variations in temperature observed in 1992 thanks to the satellite COBE (*Cosmic Background Explorer*), as a function of its direction, allowed us to know more about the violent events that preceded this phase and the subsequent formation of large galactic clusters.

6. See R. Penrose, *The Emperor's New Mind* (New York: Oxford University Press, 1989), 342.

7. See R. Lestienne, "À la mémoire de Ludwig Boltzmann: l'entropie est-elle objective?" (To the memory of Ludwig Boltzmann: is entropy objective?), *Fundamenta Scientiae* 8 (1987): 173–84.

8. P. Landsberg, "Usages et limites du concept d'entropie" (Uses and limits of the concept of entropy), *Communications* 41 (1985): 63.

9. On the other hand, the events obeying chance, in the sense that I have tried to define this term by way of the axiom of independence, are ipso facto subject to the laws of chance, such as the law of large numbers. The order associated with these laws characterizes the appearance of *order from chance.* There exists a third way to create order, of which I will speak later: *order from noise.* This last type of order is characteristic only of complex and self-organizing systems, such as living systems. They develop their complexity by taking advantage of the chances (in the exact meaning of that word) of the environment during their exchanges with the exterior.

10. R. Thom, "Halte au hasard, silence au bruit" (Halt to chance, silence to noise), in *La Querelle du déterminisme* (The quarrel of determinism), Le Débat (Paris: Gallimard, 1990): 72.

11. H. Atlan, *Entre le cristal et la fumée* (Between crystal and smoke) (Paris: Le Seuil, 1979).

12. I. Prigogine and I. Stengers, *Entre le temps et l'éternité* (Paris: Fayard, 1988), 51; English-language version, *Between Time and Eternity* (Boston: Shambala, 1989).

13. Atlan, *Entre le cristal et la fumée,* 56.

14. E. Morin, *Science avec conscience* (Science with conscience) (Paris: Fayard, 1982), 104.

CHAPTER 11: ENTROPY AND INFORMATION

1. See, for example, H. Atlan, *Entre le cristal et la fumée* (Between crystal and smoke) (Paris: Le Seuil, 1979), 64; J. Jaugh and J. Baron, *Helvetica Physica Acta* 5 (1972): 220; C. Jutten, "Mesures des entropies associées à l'influx nerveux" (Measurements of entropy related to nervous influx), unpublished thesis, Institut National Polytechnique de Grenoble, Grenoble, 1981, p. 35; G. Rossis, "Aspects épistémologiques de théories physiques utilisées dans l'interprétation des phénomènes biologiques: la théorie de l'information et le second principe" (Epistemological aspects of physical theories applied to biological

phenomena), unpublished thesis, Faculté des Sciences, Orsay, 1977; J. Tonnelat, *Thermodynamique et biologie* (Thermodynamics and biology) (Paris: Maloine, 1977), 205.

2. J. Campbell, *The Grammatical Man* (New York: Simon and Schuster 1982), 32.

3. R. Lestienne, *The Children of Time: Causality, Entropy, Becoming* (Urbana: University of Illinois Press, 1995), 114.

4. Eddington firmly believed that the growth of entropy in phenomena at our scale is the reflection of the expansion of the universe. See A. Eddington, *New Pathways in Science* (New York: MacMillan, 1935), 68. See also R. Lestienne, "Entropie, temps mécanique et flèche cosmologique" (Entropy, mechanical time, and the cosmological arrow), *Scientia* 115 (1980): 337–58. (In a celebrated comparison of very large and very small systems, Pascal used the *ciron*—a small bug—to figure a very small yet living and organized system.)

5. J. Tonnelat, *Thermodynamique et biologie* (Thermodynamics and biology), vol. 2 (Paris: Maloine, 1979), 156.

6. D. Layzer, "Quantum Mechanics, Thermodynamics, and the Strong Cosmological Principle," in *Physics as Natural Philosophy*, ed. A. Shimony and H. Feshbach (Cambridge, Mass.: MIT Press, 1982), 251. Note that the definition given by Layzer connects with Atlan's definition of redundance, or more precisely, of the product of redundancy by maximal information (Atlan, *Entre le cristal et la fumée*, 50).

7. H. Reeves, *L'Heure de s'enivrer: l'univers a-t-il un sens?* (Paris: Le Seuil, 1986), 125; English-language version, *The Hour of Our Delight: Cosmic Evolution, Order, and Complexity* (New York: W. H. Freeman, 1991).

8. D. Brooks and E. Wiley, *Evolution as Entropy*, rev. ed. (Chicago: University of Chicago Press, 1988), 346.

9. O. Costa de Beauregard, "Récents progrès dans l'analyse de l'irréversibilité physique" (Recent progress in the analysis of physical irreversibility), in *Sadi Carnot et l'essor de la thermodynamique* (Sadi Carnot and the rise of thermodynamics) (Paris: Centre National de la Recherche Scientifique, 1976), 420.

10. In 1959 Léon Brillouin, illustrating his principle of equivalence between information and entropy, showed that every physical measurement is inevitably accompanied by an increase of the entropy of the overall system constituted by the measuring instrument, the system itself, and its environment.

11. Costa de Beauregard, "Recents progrès," 49.

12. See R. Lestienne, "Charactères de la durée physique, II: mesure du temps" (Characteristics of physical duration), *Scientia* 107 (1972): 278.

13. H. Atlan, *L'Organisation biologique et la théorie de l'information* (Biological organization and the theory of information), (Paris: Hermann, 1972), chap. 9.

14. Macroinformation is defined by the natural formula

$$\bar{H} = -\Sigma_{\alpha} p_{\alpha} \log p_{\alpha},$$

copied from the usual definition of statistical entropy but applied to macrostates. Given that $H = \bar{H} + H'$, it becomes

$$H' = \Sigma_{\alpha} p_{\alpha} [-\Sigma_{i \in \alpha} p_{i/\alpha} \log(p_{i/\alpha})] = \Sigma_{\alpha} p_{\alpha} H_{\alpha} \qquad \text{(eq. 1)},$$

where $p_i = p_{i/\alpha} p_{\alpha}$. By definition, we will call H' "microscopic information."

A numerical example makes things clearer. Consider a system able to have four microscopic states, each being affected by the following probabilities: $p_1 = .1$; $p_2 = .2$; $p_3 = .3$; $p_4 = .4$. Suppose that we declare ourselves incapable of distinguishing, at the macroscopic level, between states 1 and 2 on the one hand and states 3 and 4 on the other. The macroscopic ratio α can take two values, affected by probabilities .3 and .7, respectively. At the interior of the group ($\alpha = 1$), the probabilities $p_{i/\alpha}$ take the values ⅓ and ⅔, respectively. Within the group ($\alpha = 2$), the probabilities $p_{i/\alpha}$ take the values 3⁄7 and 4⁄7. In these conditions, the formal entropy $H = -\Sigma p_k \log p_k$ has a value of 1.2799. We can easily verify that the macroscopic information has a value of 0.6109, and the microscopic information H' calculated according to equation (1) above, 0.6690. As can also be verified, we do indeed have $H' = H - \bar{H}$.

15. É. Borel, *Le Hasard* (Chance) (Paris: Alcan, 1914), 307.

16. See W. Zurek, "Atomic Randomness and Physical Entropy," *Physical Review A* 40 (1989): 4731–51.

17. See R. Lestienne, "À la mémoire de Ludwig Boltzmann: l'entropie est-elle objective?" (To the memory of Ludwig Boltzmann: is entropy objective?), *Fundamenta Scientiae* 8 (1987): 173–84. Also see Lestienne, *The Children of Time*, 122.

18. Asymmetry between the problems of prediction and those of retrodiction can easily be understood when we remember the interpretation of entropy in terms of information. In a situation where entropy is growing, information is progressively lost. The macroscopic description of the initial state remains completely sufficient. By contrast, in a situation in which entropy is decreasing, information increases progressively. The information corresponding to the macroscopic state soon becomes insufficient, and we must thus acquire microscopic information (see P. Davies, *Physics of Time Asymmetry* [London: Surrey University Press, 1974], 68).

19. T. Gold, "The Arrow of Time," in *Onzième Conseil de physique Solvay: la structure et l'évolution de l'univers* (Eleventh Solvay Council: structure and evolution of the universe) (Brussels: R. Stoops, 1958), 86–87.

20. A process is called *adiabatic* when it involves no exchange of heat with the outside. The universe considered in its entirety does not have an "exterior" with which it could exchange heat.

21. See I. Prigogine, "Talking Physics," *Nature* 304 (1983): 666.

CHAPTER 12: QUANTUM INDETERMINISM

1. The kinetic energy of a given molecule, insisted Maxwell in his debate with Quetelet, has nothing to do with the *average* kinetic energy. This latter is a collective property that has no counterpart on the scale of individual molecules, and it is the existence of this rupture that characterizes, at base, the spirit of probabilism in a break with reductionism.

2. The original formulation of the theory in terms of "de Broglie waves" is equivalent to the more modern formulation (by Dirac) in terms of "kets."

3. This anecdote was reported by E. Segré in *From X Rays to Quarks: Modern Physicists and Their Discoveries* [San Francisco: W. H. Freeman, 1980]), referring to W. Elsasser's *Memoirs of a Physicist in the Atomic Age* (New York: Sima History, 1978).

4. C. Davisson and L. Germer, "Diffraction of Electrons by a Crystal of Nickel," *Physical Review* 30 (1927): 707.

5. The principle of superposition, introduced in the wave description of matter, has epistemological consequences whose immense importance the creators of the quantum theory did not at first realize. Beginning in 1927 they hotly discussed the abandonment of classical determinism. Niels Bohr was apparently the first, in response to the famous article by Einstein, Podolsky, and Rosen (see note 6), to suggest that the properties of locality of matter are also, in no less radical a way, recalled into question. The new situation can be described in the following way: *once two quantum systems have interacted, they can never again be described by independent kets.* This is "quantum inseparability." In practice this property induces correlations between statistics of results of measurements obtained on the two systems, however distant they are from each other. This quite curious absence of independence between separate objects was the subject of numerous detailed experimental proofs, due especially to the experiments conducted at the Institut d'Optique d'Orsay by Alain Aspect at the beginning of the eighties. Since then "quantum inseparability" has been the cause of much discussion among physicists and philosophers of science (see especially on this subject the works of Bernard d'Espagnat). I will drop this question, however, for it is not directly linked to my subject.

6. A. Einstein, B. Podolsky, and N. Rosen, "Can Quantum-Mechanical Description of Physical Reality Be Considered Complete?" *Physical Review* 47 (1935): 777–80.

7. See, for example, René Thom, in *La Querelle du déterminisme* (The quarrel of determinism), Le Debat (Paris: Gallimard, 1990), 62; D. Bohm, "A Suggested Interpretation of the Quantum Theory in Terms of Hidden Variables," *Physical Review* 85 (1952): 166–93; D. Bohm, B. Hiley, and P. Kaloyerou, "An Ontological Basis for the Quantum Theory," *Physics Reports* 144 (1987): 321–75. The authors of these attempts accept the principle of superposition but seek to maintain determinism anyway. They cannot avoid the major difficulty of having to consider quantities of essentially nonlocal character.

8. The corresponding experiment is reported by A. Shimony in *Scientific American*, Jan. 1988, p. 36; see also R. Lestienne, *The Children of Time: Causality, Entropy, Becoming* (Urbana: University of Illinois Press, 1995), 92.

9. See, for example, I. Prigogine and I. Stengers, *Entre le temps et l'éternité* (Paris: Fayard, 1988), 19; English-language version, *Between Time and Eternity* (Boston: Shambala, 1989).

10. J. Wheeler and W. Zurek, eds., *Quantum Theory and Measurement* (Princeton, N.J.: Princeton University Press, 1983).

11. B. d'Espagnat, *Reality and the Physicist: Knowledge, Duration, and the Quantum World* (Cambridge: Cambridge University Press, 1989); B. d'Espagnat, *Penser la science ou les enjeux du savoir* (Thinking science, or the stakes of knowledge) (Paris: Dunod, 1990).

12. K. Popper, *The Logic of Scientific Discovery*, rev. ed. (London: Hutchinson, 1980), 248.

13. For a more in-depth discussion of this point, see M. Jammer, *The Philosophy of Quantum Mechanics* (New York: Wiley, 1974).

14. On this subject, see Lestienne, *The Children of Time*, 151.

15. Prigogine and Stengers, *Entre le temps et l'éternité*, 124.

16. K. Popper, *Quantum Theory and the Schism in Physics* (Totowa, N.J.: Rowman and Littlefield, 1982), 175.

17. B. d'Espagnat, *Une incertaine réalité* (Paris: Gauthiers-Villars, 1985), 121; English-language version, *Reality and the Physicist: Knowledge, Duration, and the Quantum World* (Cambridge: Cambridge University Press, 1989).

18. D'Espagnat, *Penser la science*, 260.

19. Recently it has been seen that mutations affecting DNA—which are essentially microscopic events, since these are sometimes pinpoint mutations due to the impact of a single cosmic ray on a single nucleotide or even on a single electron—can nevertheless have macroscopic consequences (phenotypical), insofar as they can, for example, change the physical aspect of an individual in the absence of any obvious multiplying phenomenon (the replications of DNA, according to the author of this remark, not being considered as an amplification phenomenon). See I. Percival, "Schroedinger's Quantum Cat," letter to *Nature* 351 (1991): 357.

20. W. Pauli, "Über das H-Theorem vom Answachsen der Entropie vom Standpunkt der neuen Quantenmechanik" (On the H-theorem on the increase of entropy from the point of view of the new quantum mechanics), in *Probleme der modern Physik, Arnold Sommerfeld zum 60–Geburtstage, gewidmet von seinen Schulern* (Problems of modern physics, for the sixtieth birthday of Arnold Sommerfeld, from his students) (Leipzig: Hirzel, 1928), 30–45.

21. The problem of measurement in quantum mechanics is one of the most difficult there is. As stated by d'Espagnat (*Une incertaine réalité*, 211), the quantum principle of the collapse of wave packets appears to be a process that is irreconcilable with the principles of Einsteinian relativity. The difficulty, examined especially by Aharonov and Albert, is that the reduction of the wave packet must happen on an equitemporal surface of space-time that cannot satisfy the laws of the usual transformation of relativity (Lorentz transformations) when we pass from a fixed reference frame to one that is mobile (Y. Aharonov and D. Albert, "States and Observables in Relativistic Quantum Field Theories," *Physical Review D* 21 [1980]: 3316–24; idem, "Can We Make Sense out of the Measurement Process in Relativistic Quantum Mechanics," *Physical Review D* 24 [1981]: 359–70). For a recent discussion of the problem of the relationships between quantum indeterminism and irreversibility, central in my opinion, see W. Zurek, "Decoherence and the Transition from Quantum to Classical," *Physics Today* (1991): 36–47.

22. N. Bohr, *Atomic Physics and Human Knowledge* (New York: Wiley, 1958), 20.

23. In his book *La Notion de temps* (The notion of time) (Paris: Vrin, 1983), Olivier Costa de Beauregard notes that "this possibility is the direct corollary of the complementarity between the descriptions of nuclear events in terms of interaction [due to Tomonaga-Schwinger] and the descriptions *à la Heisenberg* [which are nontemporal descriptions]; it is linked to the complementarity between numbers of occupation and phases, which is at the heart of the superquantified theory" (153; demonstration on 161). It would be necessary, however, to demonstrate more clearly that the first representation is essentially indeterministic.

CHAPTER 13: BEHIND THE COSMOLOGICAL RUPTURE LIES FUNDAMENTAL CHANCE

1. See R. Lestienne, *The Children of Time: Causality, Entropy, Becoming* (Urbana: University of Illinois Press, 1995), 122–26.

2. I. Prigogine, "The Rediscovery of Time," *Zygon* 19 (1984): 433–47.

3. R. Penrose, *The Emperor's New Mind* (New York: Oxford University Press, 1989), 328.

4. See Stephen Hawking, *Une Brève Histoire du temps* (Paris: Flammarion, 1989), 132; originally published as *A Brief History of Time: From the Big Bang to Black Holes* (New York: Bantam, 1988).

5. He was burned at the stake in Rome by the Inquisition, precisely for having suggested that the universe does not have a center. The inquisitors, naturally, did not want to deprive the Earth of the claim of supremacy that, according to them, it deserves as the celestial body that shelters humankind.

6. Lestienne, *The Children of Time,* 123.

7. Despite more and more active research on what is called, since it is not observable through our telescopes, "dark matter," no incontestable findings have yet been obtained. Theoretical arguments allow us to think that such matter must exist, however, since the motion of the galaxies cannot be explained without it. Today more and more "exotic" objects are proposed and sought: neutrinos endowed with mass (which could be as little as a few hundred-millionths of the mass of the proton); axions (of an even smaller hypothetical mass); or WIMPS (*weakly interacting massive particles*), of which the mass would be, on the contrary, much higher (on the order of one hundred times the mass of the proton).

8. Penrose, *The Emperor's New Mind,* 322, 338, 345.

9. On this subject, see R. Newton, "Probability Interpretation of Quantum Mechanics," *American Journal of Physics* 48 (1980): 1029–34. See also B. d'Espagnat, *Conceptions de la physique contemporaine* (On the conceptions of contemporary physics) (Paris: Hermann, 1965), and *Une incertaine réalité* (Paris: Gauthiers-Villars, 1985), 133; English-language version, *Reality and the Physicist: Knowledge, Duration, and the Quantum World* (Cambridge: Cambridge University Press, 1989).

Note that for the interpretation of quantum mechanics in terms of ensembles, Layzer refers to Einstein. It is true that Einstein prefered that interpretation to the objectivist interpretation of quantum mechanics, since it seemed easier to reconcile with the hope of completing quantum mechanics by the introduction of "hidden variables"; but at Einstein's time this was a debate only on principle, for no experiment could settle the question and decide between the two points of view. Today it is a *fait accompli,* thanks especially to Alain Aspect's experiments, and we know henceforth that the hypothesis of hidden variables, in its traditional form, is unacceptable.

10. See, for example, J.-P. Dufour, "La Grande Muraille de l'universe" (The great wall of the universe), *Le Monde,* Nov. 29, 1989.

CHAPTER 14: CAN WE UNMASK CHANCE?

1. In most of this chapter I will follow the remarkable article by G. Chaitin, "Randomness and Mathematical Proof," *Scientific American* 232 (May 1975): 47–52.

2. Seen through Simplicio's criteria (or Kolmogorov's), these algorithms are written at random. If they were not, it would mean that they could be compressed. This remark well illustrates the distinction that should be made between randomness or disorder on the one hand and semantics on the other.

3. Since Cantor we have distinguished several sorts of infinities, according to the numerical ensemble with which the ensemble we are considering can be biuniquely mapped. Integers have a distinct power inferior to the power of real numbers because one can interpolate an infinite number of real numbers between two integers.

4. See R. Lestienne, *The Children of Time: Causality, Entropy, Becoming* (Urbana: University of Illinois Press, 1995), 117.

5. D. Layzer, *Cosmogenesis: The Growth of Order in the Universe* (New York: Oxford University Press, 1990), 58.

6. G. Holton, *L'Imagination scientifique* (Paris: Gallimard, 1981), 227; English-language version, *The Scientific Imagination: Case Studies* (New York: Cambridge University Press, 1978).

7. A. Einstein, *Conceptions scientifiques, morales et sociales* (Scientific, ethical, and social conceptions) (Paris: Flammarion, 1952), 87.

8. K. Popper, *Un Univers de propensions: deux études sur la causalité et l'évolution* (A universe of propensities: two studies on causality and evolution) (Combas: L'Éclat, 1992), 57.

9. See, for example, A. Chalmers, "La Dépendance de l'observation par rapport à la théorie" (The dependence of observation on theory), chap. 3 of *Qu'est-ce que la science? Popper, Kuhn, Lakatos, Feyerabend* (Paris: La Découverte, 1987); originally published as *What Is This Thing Called Science? An Assessment of the Nature and Status of Science and Its Methods* (St. Lucia: University of Queensland Press, 1976).

10. Chalmers, "La Dépendance," 98.

11. See I. Lakatos, "Falsification and the Methodology of Scientific Research Programmes," in *Criticism and the Growth of Knowledge*, ed. I. Lakatos and A. Musgrave (New York: Cambridge University Press, 1974), 91–196.

CHAPTER 15: THE QUESTION OF FREEDOM

1. M. Daniel, *Montaigne et la fortune, essai sur le hasard et le langage* (Paris: Librairie Honoré Champion, 1977), 206.

2. See A. O'Hear, *Karl Popper* (London: Routledge and Kegan Paul, 1980), 146.

3. J.-P. Sartre, *L'Être et le néant*, rev. ed. (Paris: Gallimard, 1980), 495; English-language version, *Being and Nothingness: An Essay on Phenomenological Ontology* (New York: Philosophical Library, 1956).

4. Interview with H. Atlan, in G. Pessis-Pasternak, *Faut-il brûler Descartes?* (Should Descartes be burned?) (Paris: La Découverte, 1991), 55.

5. Benasayag stated that the event was essentially unforeseeable.

6. M. Benasayag, *Utopie et liberté* (Utopia and freedom) (Paris: La Découverte, 1986), 114.

7. Of disorder, Bergson wrote: "We would see what capital role the idea of *disorder* plays in the problems relative to the theory of knowledge. It does not appear there explicitly, and that is why we have not paid much attention to it. However, it is through the critic of

this idea that a theory of knowledge should begin, since if the great problem is to know why and how reality is subjected to an order, it is because the absence of any type of order appears possible or conceivable. It is that absence of order the realist and the idealist both think they are discussing—the realist when he talks about the rules that the 'objective' laws effectively impose on a possible disorder of nature, the idealist when he supposes a 'sensible diversity' that would coordinate itself—being consequently without order—under the organizing influence of our understanding" (*L'Évolution créatrice,* rev. ed. [Paris: Quadrige/PUF, 1981], 221; English-language version, *Creative Evolution* [New York: Modern Library, 1944]).

8. H. Bergson, *Essai sur les données immédiates de la conscience,* rev. ed. (Paris: Quadrige/PUF, 1988), 126; English-language version, *Time and Free Will: An Essay on the Immediate Data of Consciousness* (New York: Humanities, 1971).

9. B. d'Espagnat, *Une incertaine réalité* (Paris: Gauthiers-Villars, 1985), 195; English-language version, *Reality and the Physicist: Knowledge, Duration, and the Quantum World* (Cambridge: Cambridge University Press, 1989).

10. See P. M. Churchland, *Matter and Consciousness* (Cambridge, Mass.: MIT Press, 1988).

11. See J. Eccles, *Evolution of the Brain: Creation of the Self* (London: Routledge, 1989).

12. G. Edelman, *The Remembered Present: A Biological Theory of Consciousness* (New York: Basic, 1989).

13. P. S. Churchland, *Neurophilosophy* (Cambridge, Mass.: MIT Press, 1986), esp. 285, 317.

14. B. Libet, "Conscious versus Neural Time," *Nature* 352 (July 4, 1991): 27.

CHAPTER 16: ON THE JOY OF NOT KNOWING

1. P. de Laplace, *Essai philosophique sur les probabilités,* with a preface by R. Thom and a postface by R. Brie (Paris: Christian Bourgeois, 1986), 22.

2. A. Camus, *Le Mythe de Sisyphe,* in *Essais,* Bibliothèque de la Pleïade (Paris: Gallimard, 1965), 133.

3. A. Camus, *L'Homme révolté* (Paris: Gallimard, 1951), 375; English-language version, *The Rebel* (New York: Knopf, 1954).

4. Ibid., 129

5. Read the superb description of his breakup with Regina Olsen, his fiancée, in M. Grimault, *Kierkegaard,* Écrivains de toujours (Paris: Le Seuil, 1962), 46: "He had to send his fiancée from him; he blackened his reputation in her eyes. . . . 'She asked me: "Don't you ever want to marry?" I answered: "Maybe, in ten years, when the fire of youth has gone and I need a young girl with hot blood to make me young again." Necessary cruelty.' . . . The break was a scandal in the bourgeois circles of the capital. Kierkegaard, himself, during those nights, wept over his lost love."

6. Ibid., 82.

7. S. Kierkegaard, *The Sickness unto Death,* trans. Alastair Hannay (London: Penguin, 1968), 65.

8. S. Kierkegaard, *The Point of View for My Work As an Author,* in *Oeuvres complètes* (Paris: Éditions de l'Orante, 1971), 16:13.

9. Grimault, *Kierkegaard,* 69.

10. H. Bergson, *Les deux sources de la morale et de la religion,* rev. ed. (Paris: Presses Universitaires de France, 1990), 150–59; English-language version, *The Two Sources of Morality and Religion* (New York: Holt, 1935).

11. Ibid.

12. H. Bergson, *L'Évolution créatrice,* rev. ed. (Paris: Quadrige/PUF, 1981), 93; English-language version, *Creative Evolution* (New York: Modern Library, 1944).

13. H. Bergson, *L'Énergie spirituelle,* rev. ed. (Paris: PUF, 1972), 23; English-language version, *Mind-Energy* (Westport, Conn.: Greenwood, 1975).

14. A. Badiou, *L'Être et l'événement* (The being and the event) (Paris: Le Seuil, 1988), 432.

CONCLUSION

1. See I. Hacking's book *The Taming of Chance* (New York: Cambridge University Press, 1990), 180–89.

2. "The remarkable fact is that the new theory effectively attributes probabilities to *isolated events.* And although it considers sequences of events and the frequencies within those sequences, *the probability of an event can be completely different from its frequency in any observed segment of these sequences* (the probability of the isolated event must agree only with the *limits* of its frequency within *almost all* the sequences)" (K. Popper, *Le Réalisme et la science: post-scriptum à la logique de la découverte scientifique I* [Paris: Hermann, 1990], 393; originally published as *Realism and the Aim of Science, from the Postcript to the Logic of Scientific Discovery* [London: Hutchinson, 1988]).

3. Ibid., 384.

4. K. Popper, *Un Univers de propensions: deux études sur la causalité et l'évolution* (A universe of propensities: two studies on causality and evolution) (Combas: L'Éclat, 1992), 33.

5. Popper, *Le Réalisme et la science,* 372.

6. H. Reeves, *Patience dans l'azur* (Paris: Le Seuil, 1981), 158; English-language version, *Atoms of Silence: An Exploration of Cosmic Evolution* (Cambridge, Mass.: MIT Press, 1984).

7. Popper, *Un Univers de propensions,* 74.

8. C. Lenay, "Enquête sur le hasard dans les grandes théories biologiques de la seconde moitié du XIXe siècle" (Enquiry on chance in the great biological theories of the second half of the nineteenth century), Ph.D. diss., University of Paris-I, 1989, p. 245.

9. I. Prigogine and I. Stengers, *Entre le temps et l'éternité* (Paris: Fayard, 1988), 96; English-language version, *Between Time and Eternity* (Boston: Shambala, 1989).

10. See page 120.

11. P. Suppes, *Logique du probable* (The logic of probability) (Paris: Flammarion, 1981), 119.

12. B. d'Espagnat, *Une incertaine réalité* (Paris: Gauthiers-Villars, 1985), 207; English-language version, *Reality and the Physicist: Knowledge, Duration, and the Quantum World* (Cambridge: Cambridge University Press, 1989).

13. See J. Fetzer, "Probability and Objectivity in Deterministic and Indeterministic Situ-

ations," *Synthèse* 57 (1983): 367–86. For an opposing view, consult A. O'Hear, *Karl Popper* (London: Routledge and Kegan Paul, 1980), 135.

14. H. Bergson, *L'Évolution créatrice,* rev. ed. (Paris: Quadrige/PUF, 1981), 368; English-language version, *Creative Evolution* (New York: Modern Library, 1944).

15. J. Campbell, *The Grammatical Man* (New York: Simon and Schuster 1982), 33.

16. A. Chalmers, "La Dépendance de l'observation par rapport à la théorie" (The dependence of observation on theory), chap. 3 of *Qu'est-ce que la science? Popper, Kuhn, Lakatos, Feyerabend* (Paris: La Découverte, 1987), 162; originally published as *What Is This Thing Called Science? An Assessment of the Nature and Status of Science and Its Methods* (St. Lucia: University of Queensland Press, 1976).

APPENDIX 1

1. J. Bonitzer, *Philosophie du hasard* (The philosophy of chance) (Paris: Editions Sociales, 1984), 171.

APPENDIX 2

1. Note that if events A and B are independent, then $Pr(B/A) = Pr(B)$ and the probability of the conjugation of the events ($A \cap B$) is the product of the respective probabilities of realization of A and B.

APPENDIX 3

1. Text adapted from R. Lestienne, "Mécanique quantique" (Quantum mechanics), *Encyclopédie Alpha* (Paris: Grange-Batelière, 1971).

INDEX

RÉMY LESTIENNE is research director at France's Centre National de la Recherche Scientifique (CNRS). An elementary particle physicist, he worked with the CERN accelerator in Geneva before turning his attention toward the neurosciences. Currently pursuing his investigation of temporal mechanisms in the brain at the Institut des Neurosciences in Paris, he is also the author of *Les Fils du Temps,* published in translation as *The Children of Time* by the University of Illinois Press.

E. C. NEHER, who holds a B.A. in French from Coe College in Iowa and is the daughter of an American diplomat, was raised overseas and has studied in Paris. The coauthor of one published novel, she is currently working at the CNRS office at the Embassy of France in Washington, D.C.